19/50

D0504901

179

£6-7].

DAIRY FARM BUSINESS MANAGEMENT

DAIRY FARM BUSINESS MANAGEMENT

by

KEN SLATER

M.Sc (Newcastle), B.Sc.Ag. Hons (Dunelm)

and

GORDON THROUP

B.Sc.Hons (Leeds), Dip.Farm Man. Hons

(Seale Hayne)

FARMING PRESS LTD

Wharfedale Road, Ipswich, Suffolk, England

First published 1981

ISBN 0 85236 114 9

Printed in Great Britain by
Page Bros (Norwich) Ltd

Contents

Section 1.   **The Principles of Dairy Farm Management**

CHAPTER   *PAGE*

Preface   12

1   BUSINESS MANAGEMENT AND THE DAIRY FARMER   15

The need for a business approach—Management functions in dairy farming—Specialisation in management—Management job descriptions—Management objectives.

2   THE ECONOMICS OF DAIRY FARM MANAGEMENT   23

Basic economic objectives—Specialisation—Diversification— Competitive, supplementary and complementary enterprises—The principle of diminishing returns—Equi-marginal returns and opportunity costs—Substitution.

3   FARM ACCOUNTS AS AN AID TO MANAGEMENT   31

Historical background—Standardisation of accounting methods and management terminology—Profit definition—Account analysis— Providing the information—Comparative analysis—Fixed and variable costs—The farm balance sheet.

4   WHOLE FARM AND FIXED COST PROFITABILITY FACTORS   50

System efficiency and enterprise management efficiency—Rent equivalent—Labour costs—Power and machinery costs— Productivity per £100 labour and machinery costs—Variations in profitability between farms—Variation in profitability from year to year.

5   DAIRY HERD ENTERPRISE EFFICIENCY AND PROFITABILITY FACTORS   65

Gross margin per cow and per hectare—Milk yield—Concentrate costs—Margin over concentrates—Margin over concentrates and purchased feeds—Average milk price—Compositional quality payments—Seasonality of milk prices—Calf output—Herd depreciation (or herd maintenance cost)—Appreciation in dairy cow unit values—Sundry variable costs—Forage costs—National economic and political factors.

6   DAIRY REPLACEMENTS   88

Role in the whole farm ecomony—Profitability in the whole farm economy—Age at calving—Organising more profitable heifer rearing—Rearing for sale—pedigree breeding and rearing.

7   OTHER ENTERPRISES ON THE DAIRY FARM   100

Relative profitability—Cereals on the dairy farm—Potatoes—Sugar beet—Calf rearing—Beef cattle—Sheep—Non land using enterprises—Non-farm enterprises.

8   FARM BUSINESS ANALYSIS AND PLANNING TECHNIQUES AND THE DAIRY FARMER   112

Comparative analysis—The gross margin system—Partial budgeting—Complete enterprise costings—Gross margins as a measure of grassland efficiency—Budgetary control and management by objectives.

Section 2.   Setting up and Managing a Dairy Farm

9   RESOURCE ASSESSMENT   125

Assessing the farm and its resources—Analysis of previous year's financial performance—Assessment of overall capital position— Recurring nature of resource assessment.

10 PREPARING THE ANNUAL BUDGET AND DETAILING MANAGEMENT OBJECTIVES—A CASE STUDY 131

Resource assessment—Farming programme—Gross margin fixed cost budgets.

11 IMPLEMENTING THE FARM PLAN— MONITORING FINANCIAL PERFORMANCE 141

Monitoring MoC performance—comparing your results to other farms—Cow numbers/feed supplies—Cash flow budgeting and control—End of year managment accounts.

12 IMPLEMENTING THE DAIRY HERD PLAN—MONITORING ENTERPRISE PERFORMANCE 151

Milk yield monitoring—The brinkmanship recording system— Lactation monitoring—Effective herd recording: a case study—Feed recording—Measuring husbandry efficiency—Breeding and veterinary record.

13 CAPITAL REQUIREMENTS IN DAIRY FARMING 166

Classification of farming capital—Capital required to operate a dairy farm—Breakeven borrowed capital requirement—Capital required to operate and own a dairy farm—Classification and sources of borrowed capital.

14 THE NEW ENTRANT—GETTING A START IN DAIRY FARMING 176

Learning the job—Capital requirements and size of farm—Acquiring the capital—Acquiring the farm—Future outlook.

15 APPENDICES 183

1. Trading Account and Valuation Details
2. Terms and Definitions Used in Farm Business Management
3. Case Study Farm Schedules

INDEX 215

Tables, Schedules and Figures

Tables   *Page*
2.1. Milk price compared with cowmen's wages   24
2.2 The changes in the number of producers and cows 1955–79   25
3.1 Example of a farm trading or profit and loss account   32
3.2 Trading revenue and expenditure   33
3.3 Arriving at management and investment income   36
3.4 Mainly dairying farms: average outputs and inputs per hectare   38
3.5 Example trading account   39
3.6 Examples: (i)   gross output,
          (ii) variable costs,
          (iii) gross margin,
          (iv) fixed costs,
          (v) profit.   40–1
3.7 Calculation of individual enterprise gross margins   42
3.8 Allocation of crop variable costs   43
3.9 Change in net capital (worth) over a year   47
3.10 Statement of assets and liabilities   49
4.1 The effects of different system and enterprise management efficiencies   52
4.2 Labour costs   53
4.3 Herdsman's costs 1978 and 1980   54
4.4 Labour and machinery costs in arable and dairy farm (eastern counties)   55
4.5 Actual productivity comparisons   57
4.6 Productivity in 50 hectare-plus dairy farms   58
4.7 The variation in profitability between farms, 1979–80   59
4.8 Gross margins per cow and other livestock units, 1979–80   60
4.9 Capital structures: top and bottom 25% dairy farms, 1979–80   61
4.10 Gross margins per cow: top and bottom 25% dairy farms, 1979–80   62
4.11 Milk prices: feed costs ratios 1974–80   64
5.1 Dairy cow gross margin standards 1981   66
5.2 Analysis by yield of MMB costed farms 1979–80   67
5.3 Changes in MoC at different yield ranges   68
5.4 An analysis of ICI costed farms 1978–9   69
5.5 Example calculation of margin over concentrates   71
5.6 Rolling MoC   72
5.7 Margin over concentrates and purchased feeds   72
5.8 Butterfat classification from 1 May 1980   74
5.9 SNF classification from 1 May 1980   75
5.10 Average milk quality for the main breeds   75
5.11 Milk sales and MoCs: Friesian and Channel Island breeds   76
5.12 Average lactation yields and butterfat percentages, England and Wales,
    1978–9   76
5.13 Monthly average net prices paid to wholesale producers   77
5.14 The cost of maintaining a typical 100-cow herd   80
5.15 Utilisation of milk produced off farms in England and Wales   85
5.16 Changes in net price received for milk   86

6.1	Two examples of dairy replacement gross margins	90
6.2	Gross margins before forage costs at two ages of calving	92
6.3	Age of calving and land requirement	93
6.4	Effect on receipts and payments of earlier calving	94
6.5	Effect on gross margin of earlier calving	94
6.6	Grazing livestock gross margin budgets for a pedigree herd	98
7.1	Relative gross margins of enterprises found on dairy farms	100
7.2	Wheat enterprise gross margins	102
7.3	Output per hectare of wheat, hay and silage	103
7.4	Main variable costs of calf-rearing	106
7.5	Beef cattle margins—summer	107
7.6	Beef cattle margins—winter	107
7.7	Forecast results for autumn-born calves	108
7.8	The importance of the beef/cereal price ratio	109
8.1	Comparing gross margins and finance charges to choose between enterprises	115
8.2	Partial budget to assess effect of replacing 4 hectares wheat by 8 dairy cows	117
8.3	Separate assessment of grassland and dairy cow contributions to the total Dairy Herd Gross Margin	120
8.4	Margin over all feed at opportunity costs	121
9.1	Machinery inventory	128
10.1	The planned land utilisation	134
10.2	Estimated labour costs	136
10.3	Monthly yield per cow	137
10.4	Planning decisions and fertiliser costs	139
10.5	Average performance figures for the example farm	140
11.1	Example costs saved by selling heifers during fodder shortage	144
12.1	Milk and feed records for 21 November-calving cows	157
12.2	Silage analysis	158
12.3	Weekly feed audit sheet	162
12.4	Taking an extra 2.5 litres milk from silage	163
13.1	Capital requirement—80 hectare farm	167
13.2	Funding finance charges	168
13.3	Breakeven budget	169
13.4	Ratios between land values and dairy cow prices 1971–81	171
13.5	Annual repayment instalments for a loan of £100,000 at a fixed rate of interest of 15½%	172
13.6	The effect of changing the repayment period	173
13.7	Loan repayment methods	174
14.1	Expected profit margin on a 20–24 hectare dairy farm	177
14.2	Capital required for a 20–24 hectare dairy farm	177

Schedules (Appendix 3)

1.	Capital statement	194
2.	Gross margins and fixed costs budget summary, year ending 31.3.82	196
3.	Dairy herd monitoring data: results, year ending 31.3.81	198

4. Dairy herd monitoring data: targets, year ending 31.3.82   200
5. Grazing livestock gross margin budgets, year ending 31.3.82   202
6. Grazing livestock gross margin targets, year ending 31.3.82   203
7. Cash flow estimates, year ending 31.3.82   204
8. Notes on cash flow estimates   205
9. Accumulative cash flow estimates/results, year ending 31.3.82   206
10. Cash flow and revenue/expenditure results, year ending 31.3.82   208
11. Trading account summary, year ending 31.3.82   209
12. Gross margin accounts results summary compared to budget   210
13. Fixed costs results, year ending 31.3.82   211
14. Dairy herd gross margin, year ending 31.3.82   212
15. Youngstock gross margin, year ending 31.3.82   213

Figures
1. Management structure on a 400-hectare farm   20
2. The principle of diminishing marginal returns   27
3. Factors determining farm profits   51
4. Target growth curve: Friesian heifers calving at two years   96
5. Lactation curve for March-calving group of cows   152
6. Lactation curve for September-calving group of cows   153
7. The November-calving group of cows   156

Preface

THE SIZE of the average UK dairy herd has increased rapidly from 21 cows in 1960 to 32 in 1970 and to 57 in 1980. At the same time there has been an equally dramatic reduction in the number of milk producers. This increasing specialisation coupled with a particularly severe cost/price squeeze has brought with it a need for a greater understanding of the business and management aspects as well as the husbandry aspects of dairy farming.

A feature of dairy farming is the amount of management responsibility and business knowledge required by herdsmen and herd managers, as well as by dairy farmers themselves. This book brings together the husbandry and economic factors affecting the profitability of dairy farming, and highlights the objectives that have to be achieved if the financial result is to be satisfactory from the point of view of both the owner and the herdsman.

The authors wish to acknowledge the encouragement given them in the production of this book by ex-students and by farmer clients, some of them fulfilling both roles. We have tried to combine together our practical knowledge gained as colleagues in the field of consultancy and management with that from our joint experience in education of Houghall, Durham, and separately at Rease-Heath, Cheshire and Seale Hayne, Devon.

September 1981   KEN SLATER   GORDON THROUP
F.R.P. Consultants

Section 1

The Principles of Dairy Farm Management

Chapter 1

BUSINESS MANAGEMENT AND THE DAIRY FARMER

THE NEED FOR A BUSINESS APPROACH

Traditionally farming has been seen and treated as a way of life. In today's competitive world this is no longer feasible, except for a fortunate few; but most people still go into, or stay in, farming because they like the job, not because the 'money's good'. Most students at agricultural college are asked at some time why they want to go into farming. Many answers to the questions are given, such as:

'I like the open-air life'.
'My Father's a farmer'.
'My Grandfather was a farmer'.
'It's the only way to get free shooting and fishing'.

Only rarely is the answer *'I'm in it for the money'*, or *'It's a safe job and has a good career prospects'*.

A characteristic of farming is the relatively small size of the individual business, and in the past this has lessened the chances of an individual without capital pursuing a career in agriculture. The past twenty-five years, however, have seen a drastic reorganisation in the size and degree of specialisation in farming, not least in dairy farming, and this has opened up new job opportunities.

Part of the reason why there are more career jobs is that traditional landlords have taken 'in hand' land previously let out to tenants. There has also been the entry into farming of institutional investors. These new farmers differ from the traditional ones in their need for management expertise to run the farms and in their business approach to farming. The majority of British farmers, however, still differ from other businesses in the commercial world in the sense that they provide *both* the capital and the management expertise to run the business.

Most traditional or bona-fide farmer have also had to increase the size of their businesses to survive, and this has increased their need for specialist farm workers. This has been particularly true in dairy farming where larger herds have resulted in the need for herd managers and herdsman rather than large numbers of less skilled cowmen.

The increasing size of their farms has also led bona-fide farmers to adopt a much more business-like approach to their farming. This need for a business-like approach has been increased by the cost/price squeeze farming has undergone in recent years. As a result of these changes there have been enormous increases in productivity in farming, and this allows milk to leave the farm gate in 1980 at a price which in real terms (*i.e.*, the cost of wages) is only one-third of that twenty-five years ago.

Dairy farmers, like other businessmen, have had to learn to live with inflation and with violent fluctuations in profitability from year to year. Concentrate feed costs, for example, rose by approximately 100 per cent in an eighteen-month period during 1972–4 as a result of the world grain shortage. More recently we have seen the second oil crisis of the 1970s and its effects on the cost of energy and inflation.

Considerable farming skill and business aptitude are required to survive these various pressures in dairy farming. The number of milk producers fell from over 142,000 in 1955 to under 47,000 in 1979, a drop of 67 per cent. It seems reasonable to suppose that these trends will continue and bring with them an even greater need for a professional approach to the business of dairy farming.

This need for a professional approach to the buiness aspects of farming has to be reconciled with the typical farmer's dislike of office or paper work, and his preference for getting on with the 'proper job' of farming. This has been neatly reconciled in many instances by giving the paper work to the son or daughter recently returned to the farm from college, or to a qualified consultant/ secretary. This keeps them happy until they realise that doing the paper work is not the same as managing the business.

MANAGEMENT FUNCTIONS IN DAIRY FARMING

Management is basically about decision making, and the control and implementation of these decisions. These decisions vary in the frequency with which they have to be made and their relative importance. Some, such as whether or not to buy a piece of land, may only be made once in a lifetime, but others, such as how many concentrates or how much grass to feed the dairy cows are made every day.

A characteristic of farming—and this is particularly true of dairy farming—is the number of decisions that have to be taken by key workers. Dairy farming jobs have been mechanised and automated to a considerable extent, but most workers—and dairy herdsman in particular—still have a great deal of control, or at least potential control, of *how* a job is done. They may also decide *whether* a particular job is done and, probably most important, whether a job is done in the right way at the right time. To make the correct decision about, for example, the level of concentrate feeding to cows at differing stages in lactation, they need to understand the economic, as well as the husbandry, implications of the decision. Therefore, dairy herdsmen, as well as dairy farmers and managers, need to be capable of making and implementing decisions based on sound management principles.

Returning to the functions of the dairy farmer or farm manager: What sort of jobs does he do? What does dairy farm management involve?

Firstly, it involves making decisions to determine the policy or strategy of the business. These decisions will determine, for example, the number of cows on the farm, the number of men to be employed, the type of machines to be purchased, the method of milk production and systems of grassland production to be adopted. These decisions will also determine how the capital available is used in the business, and whether there is a need for borrowed capital or not. These decisions may be made alone, as in the case of a one-man dairy farm, or after consultation with the owner and other workers in the case of the manager of a 400-hectare estate farm. The extent to which the manager is involved in these decisions largely determines his authority and responsibilities in the business and hence the influence he can exert on the overall profitability of the business. In turn, this tends to determine the remuneration he can command.

These basic policy decisions are fundamental to the success of the farm. A large part of this book is, therefore, devoted to the economic and management principles these involve, and to the management techniques that have been developed to allow these to be put into practice.

The second stage in management is to implement the policy, or to put the plans and policies of organisation into operation. This involves many functions and occupies most of a manager's time. It is probably fair to say that most managers spend only 5 per cent of their time deciding what to do and 95 per cent doing it. The decisions they arrive at in the 5 per cent are crucial to the success of the business, but the functions undertaken in the remaining 95 per cent are equally important.

The most important area of day-to-day or enterprise management in dairy farming concerns control of the factors deciding the economic relationship or £p margin between milk sales and feed costs. This involves the control of the many husbandry factors determining milk yield per cow, and yields of forage and grassland per hectare. This control is crucial to the success of the business, and is discussed at length later in the book.

On many dairy farms husbandry control is direct, that is to say the farmer or manager himself is directly involved in milking the cows and the day-to-day management control of other enterprises. However, on the larger farm many of these day-to-day decisions become the responsibility of the enterprise manager or dairy herdsman. Consequently the job of the manager changes in emphasis, and becomes much more concerned with the control and motivation of the staff who work with him. This transition from managing crops and animals to managing people is one that many farm managers and farmers who expand their businesses find difficult to make. One of the biggest problems is the difficulty they find in delegating to others the decisions they previously took, such as deciding weekly how much concentrates to feed each cow. They no longer have at their finger-tips a whole host of pieces of information and have to pay more attention to other methods of collating this information.

This lack of contact with doing the job applies as much to the 'paper work' or office work as it does to the running of individual enterprises. On the small farm the book work only takes a few hours per month and the cash book amongst other things is written up by the farmer. On the larger farm this is not feasible, and 'administration' or control of the farm office becomes an important aspect of the manager's job. On the largest farms the term 'administrator' more nearly fits the job description of the manager than does the term farm manager, and it is fair to say that working farm managers promoted to such a position sometimes find they enjoy less job satisfaction. It needs to be stressed that one of the arts of management on the larger farm is the ability to control a job without doing it yourself. Managers often find they can retain control by doing a small part of it themselves, for instance by adding and checking the totals in the cash analysis book completed by the secretary, or by spending half an hour each week with the herd manager and/or herdsman checking lactation trends and intended feeding levels.

Communication and the need for it both up and down the lines of authority becomes increasingly important the greater the size of the farm. Fortunately in farming there are few large businesses by industrial standards and effective communications can be achieved without formal procedures. Nonetheless it remains an important function of the senior management or farm manager to see that policy decisions are communicated to all staff and acted upon. Equally it is incumbent on middle management or enterprise managers/herdsmen to inform the farm manager of the outcome of these decisions. Many decisions have to be taken with considerable uncertain knowledge of their outcome—and time or experience may prove them to be wrong. For example a decision not to feed concentrates in summer may be a policy decision on certain farms, and justified in most cases, but means need to be found whereby this decision can be changed when necessary, such as in the drought conditions of 1975 and 1976.

Buying and selling is an area of management that gives a lot of job satisfaction to farmers and managers. It is sometimes said that farmers stay in agriculture simply because it gives them the opportunity to make their weekly visits to the local market. However, to the professional manager this is regarded as largely wasted time, and he much prefers to make the decisions by telephone.

Finally a word about marketing. This is a very topical subject in farming at the present time, and in a textbook on management one would expect marketing to be rated as one of the most important functions of management. However, in dairy farming this function has been largely undertaken for nearly fifty years by the Milk Marketing Board. Consequently there has been little need for the individual farmer to spend time marketing his product and, despite the publicity on this topic, this is likely to continue to be the case in the future. This does not mean that the dairy farm manager should ignore marketing. He needs to be aware, for example, of the reasons why the Milk Marketing Board has recently changed its method of payment for milk quality. He should be examining his business to see what changes, if any, are necessary in his production methods so that he can achieve the best profit from his farm under the present pricing policy.

SPECIALISATION IN MANAGEMENT

The fact that most farm businesses are small in industrial terms has already been mentioned, and this means that there is relatively little specialisation in management. One rarely finds financial managers, sales managers, personnel managers, works managers etc., each with their specialist functions. On the majority of farms all these functions have to be performed by the same person, and on the smallest farm—the one-man farm—the manager is also the work force!

It has, however, to be accepted that the size of many farm businesses is growing, and thought has to be given to what form of management structure to adopt.

The management structure on a 400-hectare mainly dairy farm keeping 450 dairy cows, 400 young stock and growing 300 hectares of forage crops and 100 hectares of cereals could be on the lines shown in fig. 1.

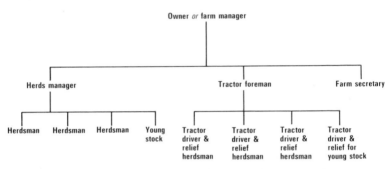

Fig. 1. Management structure on a 400-hectare farm

On such a farm it is necessary to be clear in defining the responsibilities and authority that goes with each individual job, so that each person knows what is expected from him or her.

In the above chart it is assumed that all the tractor drivers act as relief stockmen, but on many farms this would not be feasible. Consequently the herds manager may find that one of his main jobs is to act as a relief milker to the herdsman under his control. This will probably be necessary anyway if he is to be gainfully employed because there is not sufficient 'management work' to justify both his and the farm manager's employment. This will depend, however, to some extent on the nature and inclinations of the manager, as well the capabilities of the tractor foreman and whether the latter is simply a foreman or an arable manager capable of taking more responsibility for decisions such as which spray to use as well as when to apply it. Given the above management structure there will be relatively little management responsibility for the herdsmen, and this title is possibly a misnomer. It would be probably more accurate to use the rather more old-fashioned term cowmen, because most of the time they would be carrying out work under close supervision.

MANAGEMENT JOB DESCRIPTIONS

Given the above structure, the job description and management responsibilities of the farm manager and herds manager would be along the lines shown below.

Farms Manager

Responsible for the implementation of the farm business policy and strategy agreed at regular meetings with the owner and his advisers. He is accountable directly to the owner.

He is required to liaise closely with the Estate Land Agent and Estate Foreman.

The duties of the farms manager include:
(a) The preparation for approval annually of a profitable working, cropping, stocking and capital programme together with appropriate financial budgets.
(b) The purchase and sale of such goods and services as are required to implement the agreed policy.
(c) The day-to-day control and organisation of farm staff. Responsibility for the recruitment and dismissal of farm staff according to policy and after consultation with the owner.
(d) The day-to-day control of all farming operations in accordance with the agreed policy.
(e) The production and keeping of such records as may be required for the profitable functioning of the farm.
(f) The keeping of farm machinery and equipment and the maintenance of farm buildings and premises in a tidy condition.
(g) The supervision of capital improvements as and when agreed as part of the policy.

Herds Manager

The Herds Manager should:
(a) Prepare annually for approval a dairy herd farming programme including proposed culling and replacement policies, gross margin budgets for each individual herd, and capital expenditure proposals.

(b) Implement the approved dairy herd farming programme under the over-all control and supervision of the farms manager.

(c) Accept responsibility for keeping all records pertaining to the dairy herd.

(d) Supervise all staff employed on a permanent or temporary basis to look after the dairy cows and young stock.

(e) Liaise closely with the tractor foreman in all matters relating to the management of the dairy herd and grassland.

MANAGEMENT OBJECTIVES

An agricultural economist would define the main objective as being to maximise the profit that could be obtained from the business. In practice, however, there are many differing objectives to satisfy, some of which conflict with the maximisation of profit.

The most important factor to consider is one's own personal objective in relation to the business in which one is working. This means asking oneself three questions, and these three questions are the same whether you are the herdsman, the farms manager, or the owner-manager. They are:

1. What am I trying to do?

2. What is stopping me from doing it?

3. What can I do about it?

They are essential to the job of management, whether you are considering the long-term future of your business, or trying to decide what to do next week, or trying to decide on the goals in your life. The purposes of this book is to try to help you to put and answer these questions because basically that is all that management or life are about.

Chapter 2

THE ECONOMICS OF DAIRY FARM MANAGEMENT

BASIC ECONOMIC OBJECTIVES

The basic economic objective in dairy farming, as in the management of any farm, is to combine together land, labour and capital in such a way as to maximise profit, or at least to obtain a substantial profit over the long term.

To achieve this economic objective the farmer or manager has first to decide what to do and secondly to do it. This chapter is concerned primarily with the economic principles underlying the decision what to do; or, to put it more professionally, deciding on the overall plan and objectives of the business. To arrive at a farm policy, management has to decide:

1. (a) Which enterprises to have on the farm;
   (b) how large each enterprise should be.
2. (a) The method of production to adopt for each enterprise;
   (b) the yield at which to aim.

The economic criterion used to arrive at these decisions is, 'What system to adopt to maximise the profit from the *whole farm*', not just from the *dairy herd*.

The main economic problems involved in these decisions are the degree of specialisation and intensity of production to adopt in running the farm.

Relatively few farms are devoted one hundred per cent to milk production. The objective in the first part of this chapter is to consider some of the reasons why, and at the same time to consider the economic principles and other factors that need to be taken into account when deciding on the number and size of each individual enterprise.

A second feature of dairy farming is the wide variations found in methods of production and yield levels. The second part of the chapter considers the economic and other factors to be taken into account in deciding what system of production to adopt and the yield level at which to aim.

23

SPECIALISATION

Having said that very few farms are devoted one hundred per cent to milk production it is fair to say that dairy farms are among the most specialised in British agriculture.

Specialisation in milk production is feasible because dairy farming can be practised over a wide range of soil types and there are fewer technical problems to overcome compared, for example, to specialised cereal farms. A feature of farming in recent years has been the increased specialisation that has taken place in all farming systems and this has been particularly true in dairy farming. The reasons for this are numerous and the more significant are itemised below.

- The need to increase cow numbers to make effective use of improved housing, feeding and milking systems.
- The technological knowledge required has increased considerably in all enterprises. An advantage of specialisation in dairy farming, as in other enterprises, is the opportunity it affords to come to terms with this technology and to implement it effectively.
- The need to eliminate less profitable enterprises to overcome problems generated by the cost/price squeeze, that is, the decrease in product prices relative to costs that has continued unabated for the last twenty-five to thirty years.

Table 2.1. Milk price compared with cowmen's wages

Year	Milk price (pence per litre)	Weekly average total earning and hours of hired, regular wholetime adult male dairy cowmen in Great Britain*		No. litres equivalent to weekly earnings
		£	hours	
1954–5	3·326	8.72	57·1	262
1959–60	3·220	11.64	56·9	361
1964–5	3·403	15.15	56·4	445
1969–70	3·548	20.63	54·8	581
1974–5	6·245	45.03	53·2	721
1978–9	10·430	78.64	53·4	754
1979–80	11·40†	90†	53	789†
1980–1	12·50†	112†	53	896†

* Source MAFF, DAFS
† Estimates

An indication of the cost/price squeeze that the dairy industry has undergone can be seen from an examination of Table 2.1. 754 litres of milk were needed to pay one man's weekly wage in 1978–9 compared to 262 litres in 1954–5, an increase of 188 per cent. This trend is expected to continue, and at least 1,040 litres will probably be required by 1984–5, that is at least four times the quantity required thirty years earlier.

The trend towards greater specialisation in dairy farming is illustrated by Table 2.2. During the period 1955 to 1979 the number of producers fell from 142,792 to 46,972 or by 67 per cent and at the same time the size of the average herd increased from 17 to 58, or by 241 per cent.

Table 2.2. The changes in the number of producers and cows 1955–79

Year	No. producers	No. dairy cows (000)	No. cows per producer	Yield per cow (litres)
1955	142,792	2,415	17	3,065
1960	123,137	2,595	21	3,320
1965	100,449	2,650	26	2,545
1970	80,265	2,714	34	3,755
1975	60,279	2,701	45	4,070
1979	46,972	2,727	58	4,745
1979 as per cent 1955	33	113	340	155

DIVERSIFICATION

Despite the trend towards specialisation there are still good reasons why many dairy farms do not concentrate solely on milk production.

1. Lack of adequate buildings and/or lack of land suitable for dairy cows are major reasons. The former may only be a temporary restraint and can be overcome given adequate capital.

2. Problems encountered in disposing of slurry restrict the number of cows that can be kept and may lead to the introduction of an enterprise such as spring cereals.

3. Many farms remain diversified simply because the farmer likes it that way even if this may mean a lower overall profit at the end of the day. This is particularly true when it comes to rearing dairy replacements.

4. There is a case for diversification to reduce risks and to even out profits from year to year. The risk of introducing disease is

regarded by many farmers as a good reason for rearing young stock.

Feed prices vary considerably from year to year relative to milk price and this leads to considerable fluctuation in profits. Growing cereals is a way of reducing the effect of these price changes and consequent variations in profits from year to year.

5. The success or otherwise of many businesses, not only in farming, depends on the efficient disposal of by-products. The calf is a major by-product of the dairy farm and its effective disposal may necessitate or be an added reason why a beef enterprise should be introduced.

6. Surplus labour at certain times of the year or all the year may lead to the introduction of a new enterprise such as calf rearing or a small pig enterprise so as to justify maintaining this labour on the farm.

7. The ability of the land to grow highly profitable crops such as potatoes and winter wheat will also lead to diversification. In certain parts of the country the introduction of such a crop is particularly advantageous as it allows double cropping of the land, for example early potatoes followed by kale or stubble turnips.

COMPETITIVE, SUPPLEMENTARY AND COMPLEMENTARY ENTERPRISES

The discussion to date has centred round the reasons for specialisation and diversification. In practice, it is up to the individual farmer to decide on the right balance of enterprises at his disposal. In this connection it is usual to designate enterprises as being competitive, supplementary or complementary to each other in relation to their demands for land, labour and capital.

A calf-rearing enterprise on a dairy farm is competitive to the dairy cows in the need for capital but non-competitive for land. It may be feasible to carry this enterprise without any additional labour and utilise buildings which would otherwise remain idle; *i.e.*, it is a supplementary enterprise so far as labour and buildings are concerned. On many dairy farms there is an area of land not suited to grazing by dairy cows and this can be grazed by dairy heifers *without* competition to the dairy herd. Whether or not this competition exists between enterprises, whether it be for land, labour or capital needs to be assessed most carefully when deciding on the cropping and stocking of the farm.

An enterprise is said to be complementary when it makes a contribution to the success of the other. The rearing of dairy heifers contributes to the success of a dairy herd *if* heifers reared perform better than purchased replacements. Some of the merits of ley farming are questioned but most people would expect crops such as potatoes and winter wheat to yield better than normal quantities following a ley and/or to be grown at less cost due to savings in fertilisers. In this case the dairy herd can be said to have a complementary effect on the cereal crops.

One of the arts of management is to be able to devise a farming system that exploits the complementary and supplementary relationships between enterprises and minimises direct competition for resources. This applies whether one is considering the role of cereals on a dairy farm or the place of dairy cows on an arable farm. The term 'art of management' is used because it is very difficult to quantify these relationships and express them as £p.

THE PRINCIPLE OF DIMINISHING RETURNS

This principle has special application when we are considering the yield level at which to aim whether it be milk yield per cow or grass yield per hectare.

We can expect to increase the yield per cow by providing greater and greater quanties of high energy or concentrate feed. Similarly the yield of grass can be increased by using greater and greater quantities of nitrogen. However, a point is reached when additional inputs tend to produce a diminishing increment of output for each additional input. Eventually a point is reached where additional input leads to loss of output. This relationship is illustrated in the graph below:

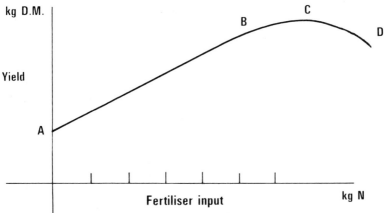

Fig. 2. The principle of diminishing marginal returns

Between A and B there is a linear relationship, *i.e.*, the increase in yield for a given additional input is constant. Between points B and C there is a diminishing return for additional inputs and yield eventually falls between points C and D.

The *maximum* yield occurs at point C but the *optimum* yield occurs at a point between B and C. At this point the value of the additional or *marginal* increment of output is equal to the value of the additional or *marginal* increment of input. It is also the point at which the margin of yield over inputs is at its maximum.

It is relatively easy to state this principle but in practice it is difficult to evaluate when the optimum point occurs due to the complex technology of the dairy farm enterprise. Hence the conflicting arguments regarding the optimum level of concentrate feeding to dairy cows. One school of thought will put the optimum at less than one tonne per cow per annum with yields in the region of 5,000 litres per cow, representing a concentrate use of less than 0.20 kg per litre, whereas another will put the optimum level of feeding at 0.35 kg per litre with expected yields in excess of 6,500 litres.

A similar case is nitrogen inputs on grassland. It is generally accepted that a linear relationship exists between yield and nitrogen input up to about 300 kg per hectare. Considerable conflict of opinion, however, occurs when it comes to the effect of frequency of cutting on yield. Fewer cuts will increase the total yield of dry matter but reduce the quality of grass conserved. In this instance it is very difficult to evaluate the £p value of the marginal output of, say, silage for the additional marginal input of labour and machinery.

This need for additional inputs of labour and machinery may also occur at the same time as it is needed for another enterprise; for example, third-cut silage often clashes with the cereal harvest. The problem to be answered in this case becomes the wider one of: 'Should we use our labour and machinery to harvest silage or to harvest cereals?'

THE PRINCIPLE OF EQUI-MARGINAL RETURNS AND OPPORTUNITY COSTS

To answer the question posed in the last paragraph we need to work out the value of the third-cut silage less the cost of making it, and compare this to the cost of having the cereals combined by a contractor and the adverse effect this may have on the value of the cereal harvest. As a result we would hope to arrive at a decision that maximised the profit from the farm as a whole but in doing so we would have to accept a lower net return from one of the enterprises. This lower return from the one enterprise would represent the *opportunity cost* of making a profit from the other. Some economists would argue that opportunity costs are the only real costs. For example, the *real* cost of growing a silage crop on an arable farm is the margin that could be obtained from growing a cereal crop.

This principle of opportunity costs is fundamental to the approach we should make in our decision making in farm management. Partial budgets and gross margin budgets represent methods of application of this principle and are discussed in more detail in Chapter 8. They are the means whereby we hope to achieve the optimum balance of enterprises and methods of production on the farm. This situation will attain when the marginal returns from all enterprises are equal.

PRINCIPLES OF SUBSTITUTION

When considering the principle of diminishing returns we simply considered the effect at the margin of using more and more of an input such as nitrogen or concentrates in relation to the yield from a given area of land or a dairy cow. However, we also need to consider whether one product can be substituted for another to produce a given yield, the objective being to substitute a low-cost input such as grazing for concentrates.

In the feeding of a dairy cow we are not simply considering the effect of changes in the amount of concentrates fed. We are also changing the amount of other feeds fed and *substituting* silage or other bulk feeds for concentrates and vice versa. This principle is forgotten by many so-called experts who draw erroneous conclusions from figures showing the relationship between milk yields and concentrate inputs that assume 'other things are equal', whereas in fact they are quite different.

To date, much of the discussion has centred round inputs such as feed and fertilisers but the substitution of capital in terms of machinery, equipment and buildings for labour is of equal if not greater importance. Again, however, this cannot be considered in isolation from the rest of the farm. It may be profitable to spend capital on machinery and equipment to speed up the rate of silage making but it might be more profitable to spend it on more cows and/or on improving the milking facilities.

This takes us back once again to the principle of equi-marginal returns and opportunity costs. This should be central to our decision-making process in the preparation of our plans and policies for the farm.

If we understand and use this principle effectively we will achieve a good balance of enterprises and a balanced method of producing the main product, *i.e.*, milk from our farm. Whatever the problem and whatever the challenge we need to ask questions such as 'Is there a better, less costly way of doing this job?' 'Do I need to do the job now or would it be better to do something else?'

However, we must not spend all our time thinking what to do or how to do it; that is the privilege of academics! We have to make a decision, get on and do the job or tell someone else to do it. Later however, we must not forget to check the job was done in the way intended and yielded the result expected. This will make the decision much easier the next time.

Chapter 3

FARM ACCOUNTS AS AN AID TO MANAGEMENT

HISTORICAL BACKGROUND

Prior to the 1939–45 war farmers did not need to prepare accounts for tax purposes, because income tax was charged on a per hectare basis. During the war this system came to an end and farmers became obliged to prepare tax accounts duly drawn up and certified as being correct by a fully qualified accountant.

This resulted in the development during the post-war years of accounting methods which were only designed to supply information for taxation purpose and these were of little or no use for management. The need to keep these accounts for tax purposes also engendered a deep suspicion and hatred of accounting in the farming community.

Farm accounts and records was taught in colleges for many years as a subject which was an end in itself, and as such it was unpopular with staff and students alike. Before the introduction of farm business management techniques in the late 1950s and early 60s it was not realised that accounts could and should provide the material for a detailed study of the dairy farming business.

The growing awareness of the value of accounts as an aid to management was largely the result of the work of the farm management specialists appointed to the staffs of the agricultural economics departments of universities responsible for collecting financial information from farms for Price Review purposes. Each year these universities produce financial data which is used to establish the average net farm incomes of British farmers. This formed an essential part of the Annual Price Review Negotiations for many years and it is still the main source of financial data for the Government in its Common Market negotiations.

The data provided by the agricultural economics departments is fully authenticated and provides an undisputed source of reliable information on the economics of dairy farm production and other farm enterprises. The information is derived from accounts collected from bona-fide farms all over the country.

31

In the late 1960s the Ministry of Agriculture introduced a Farm Business Recording Scheme. Under this scheme grants were paid to farmers for the production of management accounts providing these were presented in a standardised form. The standardised trading account and other data that had to be produced to qualify for grant is shown in Appendix 1. This grant aid scheme did much to stimulate the use of accounts as an aid to management. Colleges set up courses to train farmers and farm secretaries to produce accounts suitable for management purposes and this work has continued to expand and develop over recent years. Commercial organisations have set up management costing schemes and foremost amongst these from a dairy farmer's point of view has been the work of the Milk Marketing Board's Farm Management Services Department.

STANDARDISATION OF ACCOUNTING METHODS AND MANAGEMENT TERMINOLOGY

Over the years it is fair to say that advisers and farmers have come to realise that farmers need two sets of accounts, one for tax purposes and one for management purposes. This need arises from the need to compare 'like with like' when comparing an individual farmer's results to those achieved by other farmers. It is also important to use the same terminology if confusion is not to arise when communicating with other farmers and advisers. For these reasons the Ministry of Agriculture has produced a *Glossary of Terms and Definitions Used in Farm Management*, and this is reproduced in Appendix 2.

Before discussing these various terms and definitions we need to consider the format of a farm trading or profit and loss account. A simple example is shown in Table 3.1 and a more detailed example is shown in Appendix 1.

Table 3.1. Example of farm trading or profit and loss account

	£		£
Opening valuation	95,000	Closing valuation	100,000
		Trading revenue	
Trading expenditure	130,000	(or income)	160,000
Depreciation	6,000	Notional income	1,000
Profit	25,000	Loss	—
	261,000		261,000

This account shows a profit of £30,000 after charging £6,000 depreciation and including £1,000 notional income. The latter represents the value to the farmer of the produce his family has consumed during the year such as milk, and also takes into account the rental value of the farmhouse and the value of other private consumption items charged to the business during the year.

In the example the trading revenue exceeds the trading expenditure by £30,000, that is £160,000 less £130,000. This surplus of trading revenue over expenditure is not the same as the trading *cash* surplus because account has to be taken of the changes in creditors and debtors. How revenue and expenditure relate to cash flow is illustrated in Table 3.2.

Table 3.2. Trading revenue and expenditure

	£		£
Trading payments	135,000	Trading receipts	159,000
Add closing creditors	10,000	Add closing debtors	16,000
	145,000		175,000
Subtract opening creditors	15,000	Subtract opening debtors	15,000
TRADING EXPENDITURE	130,000	TRADING REVENUE	160,000

The surplus of trading revenue over expenditure is £30,000 but the trading *cash* surplus, that is trading receipts less trading payments is only £24,000. The cash surplus is £6,000 less than the revenue/expenditure surplus because creditors have been reduced by £5,000 and debtors have increased by £1,000.

It is important to distinguish and to remember the difference between *receipts* and *payments* on the one hand, and *revenue* and *expenditure* on the other.

When preparing farm accounts it is also most important to appreciate the enormous effect that *subjective* judgements made regarding the valuations of livestock and other commodities on hand at the beginning and end of the year can and do have on the profit. The same relative values for a commodity should be used at both the beginning and end of the year unless there has been a *real* change in values so as to avoid showing profits (or losses) that are simply due to differences in valuation judgements. On dairy farms it is also important to avoid showing wide fluctuations in profits (or losses) that are simply due to changes in the value

of home-grown forage supplies on hand between the beginning
and end of the year.

In the trading account example used earlier the valuation has
increased by £5,000, that is from £95,000 to £100,000. This could
be simply due to an increase of £50 per head in the value of one
hundred dairy cows, the numbers on hand being the same at the
beginning and end of the year, or it could represent an extra ten
cows at £400 per head, plus an extra five heifers at £200 per head.

When analysing a farmer's financial results it is necessary to look
thoroughly at the valuation details to see if they are masking the
true picture. Unfortunately, this is often not feasible, particularly
if the only information available is the income tax accounts, as the
valuation produced by the valuer may have been arranged to
produce a given result rather than vice-versa.

Valuations for tax purposes are usually produced on the basis
of cost of production or two-thirds market value. Valuations for
management purposes need to be produced on estimated market
values if they are to show the true result. Valuations on this basis
are not desirable from a taxation point of view so in most instances
it is necessary to have two valuations: one for tax purposes and
one for management purposes.

Finally, before leaving valuations a few comments need to be
made on dairy cow values and the impact inflation has had on
these values. At the present time (1981) a newly-calved cow/heifer
entering a herd has a value in the region of £500–£550 whereas the
average cull cow (including casualties) has an average value in the
region of £300–£350. A figure midway between these two, *i.e.*, in
the region of £400–£450 per head, is therefore an appropriate value
for a herd at the present time. Ten years ago the figure calculated
on the same basis would have been in the region of £100. Just one
year ago the figure would have been in the region of £350–60.

Dairy cow values have tended to keep up with inflation over the
years and it is important to show the effect of this 'hedge against
inflation' in the management accounts. If this inflation hedge is
not included the relative profitability of dairy farming compared
to, say, arable farming is underestimated. When the final management accounts are prepared this 'appreciation' in unit stock
values should be shown as a separate item.

Having discussed the problems of livestock values and inflation
we need to look briefly at the problems inflation has brought in
relation to the calculation of depreciation. Traditionally, machinery
depreciation has been calculated on an historic cost basis, usually
on a reducing balance method. For example:

	£
Tractor purchased 1978	8,000
Depreciation 1st year 25 per cent	2,000
Written down value 1979	6,000
Depreciation 2nd year 25 per cent	1,500
Written down value 1980	4,500
Depreciation 3rd year 25 per cent	1,125
Written down value 1981	3,375
Accumulated Depreciation	4,625
Initial Purchase Price	8,000

Depreciation is meant to cover the cost of replacing the machine when this becomes necessary. The accumulated historic cost depreciation of the tractor in the above example is £4,625, and its written down value is £3,375. The purchase price of an equivalent replacement in 1981 is likely to be in the region of £12,000 and the trade-in value of the old tractor is likely to be in the region of £5,000. The old tractor is undervalued in the books by £1,625 and 'tax' would be payable on this paper profit if a decision was taken to give up farming and sell the tractor. The accumulated depreciation, on the other hand, of £4,625 is £2,375 less than the £7,000 (£12,000 − £5,000) required to replace the tractor if a decision is taken to continue in farming.

There is no simple solution to this problem. Cambridge University, for example, now produce their reports with depreciation calculated by both historic cost and current cost methods. Whichever method is used the result shown is very subjective and open to error. The only really worthwhile advice that can be given is to treat all depreciation figures with caution and to accept that at the best the depreciation figure is only an estimate. Note should also be taken that depreciation figures produced on the historic cost basis will grossly underestimate the capital sum required to meet current machinery replacement requirements.

PROFIT DEFINITION

Having discussed some of the principles involved in the preparation of a farm trading or profit and loss account we now turn to the adjustments that need to be made to the accounts so that 'profit' can be accurately defined and standardised so that the results can be compared to those achieved by other farms. The basic objective in making these adjustments is to arrive at the Net Farm Income which is defined as the 'return to the farmer and his wife for their manual labour, management and tenants' capital invested in the business'. Any interest charges they pay are deducted; if they are owner-occupiers a rented value is included for the land owned and ownership expenses incurred in respect of this land are excluded. Net Farm Income is defined as the return to the 'farmer and his wife' only, so to arrive at this figure the value of work done by unpaid family labour has to be added to the actual paid wages. Finally any non-farm expenses are extracted from the accounts and an addition is made for the value of *notional benefits* the farmer and his wife receive, such as the rented value of the farmhouse, the value of produce consumed, the value of electricity and fuel consumed in the house, and costs saved by having the personal use of the farm car. These non-farm costs and notional benefits can be substantial and, on the smaller farm in particular, they often represent a large proportion of the total net farm income. At the best these estimates are only approximations. Consequently it is often difficult to ascertain the true profit made

Table 3.3. Arriving at management and investment income

	£ total	£ per hectare
Profit according to income tax accounts	15,000	250
add		
Interest charges paid to bank	1,200	20
Ownership expenses	300	5
Notional income	1,200	20
	17,700	295
subtract		
Rental value	6,000	100
Value of son's unpaid manual labour (including 14% national insurance)	6,600	110
Value of farmer's unpaid manual labour	6,600	110
MANAGEMENT AND IVESTMENT INCOME	− (1,500)	(25)

by a small farm, and the true costs, particularly power and machinery costs, involved in running a farm are difficult to gauge due to the effect on these of private cars and fuel consumption.

Having calculated Net Farm Income it is usual to substract the value of the farmer's and wife's *manual labour* in order to arrive at the Management and Investment Income, which represents the reward to management and return on tenant's capital invested in the farm. It is unfortunate that this is the final approved definition of profit as this still measures the return before making a charge for management. Many farm businesses are now run by salaried managers and it would be helpful to have standard 'Investment Income' data.

The salutory effect the changes outlined above can have on the 'profit' made by the farm is illustrated in Table 3.3. The farmer concerned is an owner-occupier on sixty hectares. He is reasonably complacent because he and his son who is a partner in the buiness have made a 'profit' according to his income tax account of £15,000. However, after making the adjustments referred to above, we find that this is reduced to a Management and Investment Income of (−) £1,500.

ACCOUNT ANALYSIS

So far our discussion has concerned the various ajustments we need to make to the farm accounts and the care that has to be taken in their preparation if one is to be able to compare the profit made by one farm to another and draw sensible conclusions as to their relative profitability.

Universities and other organisations produce standard performance data and the results shown in Table 3.4 are for a group of dairy farms costed by Manchester University. If we look at this data we see immediately that they are not presented in the form of a conventional trading account as described earlier in this chapter, but are in terms of Gross Output, Variable Inputs (Costs), Gross Margin, and Fixed Inputs (Costs) per hectare. If we look at the results more carefully we will also notice that the profit or Management and Investment Income equals the Gross Margin less the Fixed Inputs, and that the Gross Margin equals the Gross Output less the Variable Inputs.

To explain these terms it is now proposed to take a relatively simple example trading account for a 40 hectare dairy farm and

Table 3.4. Mainly Dairying Farms: Average ouputs and inputs per hectare

		Over 50 hectares		
		1978–9	1979–80	
		Average £	Average £	High profit £
Gross Output per Hectare				
Main crops*		55·5	55·7	67·2
Forage crops and main crop by products†		6·7	3·8	3·8
Milk		765·6	878·7	970·4
Cattle		275·1	210·9	240·2
Sheep and wool		9·2	6·5	9·7
Pigs		6·8	5·3	11·8
Poultry and eggs		0·1	0·2	—
Miscellaneous		8·8	10·8	16·3
TOTAL GROSS OUTPUT	(1)	1,127·8	1,171·9	1,319·4
Variable Inputs per Hectare				
Feed: purchased		322·8	424·5	413·0
home produced**		30·9	32·2	39·0
Seeds: purchased		11·1	10·0	8·4
home grown		0·8	0·5	—
Fertilisers		70·3	78·2	84·8
Other inputs: crops		9·2	10·8	10·9
livestock		42·6	50·2	50·0
TOTAL VARIABLE INPUTS	(2)	487·7	606·4	606·1
GROSS MARGIN PER HECTARE {(1)−(2)}	(3)	640·1	565·5	713·3
Fixed Inputs per Hectare				
Power		159·4	187·1	187·7
Labour inc. farmer and wife		143·5	166·1	154·7
Land expenses: repairs and maintenance		15·9	15·9	13·8
rent and rates		64·8	78·4	81·8
General overhead costs		19·8	27·0	22·5
TOTAL FIXED INPUTS	(4)	403·4	474·5	460·5
MANAGEMENT AND INVESTMENT INCOME {(3)−(4)}	(5)	236·7	91·0	252·8
Farmer's and wife's labour	(6)	43·3	48·7	58·4
NET FARM INCOME {(5)+(6)}	(7)	280·0	139·7	311·2
Breeding livestock appreciation (BLSA)		105·6	33·8	33·6
NET FARM INCOME LESS BLSA		174·4	106·6	277·6

* Includes Farm Consumed
† Includes only sales and valuation
** Includes bucket-fed whole milk
(Farm Management Survey, Department of Agricultural Economics, University of Manchester.)

show how this is analysed to produce gross outputs, fixed costs, variable costs and gross margins. Having read this section the reader is then recommended to study the schedules for the Case-Study Farm, shown in Appendix 3, and to read the Glossary of Terms in Appendix 2.

Table 3.5 Example trading account

Opening Valuation	£	Closing Valuation	£
Dairy cows: 60	24,000	Dairy cows: 64	25,600
Dairy replacements: 40	10,000	Dairy replacements: 42	10,500
Hay and silage	1,000	Hay and silage	1,200
Seeds	200	Seeds	Nil
Fertilisers	1,000	Fertilisers	6,000
Purchased feed	1,000	Purchased feed	600
Fuel and oil	200	Fuel and oil	100
TOTAL	37,400	TOTAL	44,000

Expenditure	£	Revenue	£
Dairy cows	1,000	Milk	40,300
Feedingstuffs	15,000	Cull cows	7,400
Vets and medicines	1,200	Calves	1,600
A.I. & dairy expenses	1,000		
Other livestock expenses	600		
Seeds	200		
Fertilisers and lime	10,000		
Sprays and other crop expenses	500		
Wages and national insurance	5,000		
Fuel and oil	1,000		
Electricity	1,000		
Machinery repairs, tax and insurance	2,500		
Contract	600		
Rent	3,200		
Rates	300		
Property repairs	1,000		
General insurance	500		
Professional fees and office expenses	1,000		
Sundry overheads	500		
TOTAL	46,100	TOTAL	49,300

Depreciation		Notional Income	
Machinery and equipment	5,000	Milk	300
Tenant's improvements	800	Rental value of house	500
		Use of car	800
	5,800		1,600

PROFIT	5,600		
	94,900		94,900

The Gross Output, Variable Costs, Gross Margin and Fixed Costs
are determined as shown in Table 3.6.

Table 3.6. Example (i). Gross output	£	£	£
(a) Milk: revenue (sales)		40,300	
plus milk consumed		300	40,600
(b) Cattle: cull cow sales	7,400		
calf sales	1,600	9,000	
plus closing valuation: dairy cows	25,600		
dairy replacements	+10,500	36,100	
		45,100	
less opening valuation: dairy cows	24,000		
dairy replacements	+10,000	34,000	
		11,100	
less dairy cow purchases		1,000	10,100
(c) Hay and silage: closing valuation		1,200	
less opening valuation		1,000	200
(d) Notional Income: Rental value of house		500	
Use of car		800	1,300
TOTAL GROSS OUTPUT			52,200

Table 3.6 (ii). Variable costs	£	£
(a) Feedingstuffs: expenditure	15,000	
plus opening valuation	1,000	
less closing valuation	600	15,400
(b) Livestock sundries: vet and medicines	1,200	
plus A.I. and dairy expenses	1,000	
other livestock expenses	600	2,800
(c) Seeds: expenditure	200	
plus opening valuation	200	
less closing valuation	Nil	400
(d) Fertiliser and lime expenditure	10,000	
plus opening valuation	1,000	
less closing valuation	6,000	5,000
(e) Other crop expenses	500	
+ Opening valuation	Nil	
− Closing valuation	Nil	500
TOTAL VARIABLE COSTS		24,100

Table 3.6 (iii). Total gross margin

	£
Total gross output	52,200
less total variable costs	24,100
TOTAL GROSS MARGIN	28,100

Table 3.6 (iv). Fixed costs

	£	£
(a) Labour: wages and national insurance	5,000	5,000
(b) Power and machinery: fuel and oil expenditure	1,000	
plus opening valuation	200	
less closing valuation	100	
	1,100	
electricity	1,000	
machinery repairs, tax and insurance	2,500	
contract	600	
machinery and equipment depreciation	5,000	10,200
(c) Property charges: rent	3,200	
rates	300	
property repairs	1,000	
tenants' improvements depreciation	800	5,300
(d) General overheads: general insurance	500	
professional fees and office expenses	1,000	
sundry overheads	500	2,000
TOTAL FIXED COSTS		22,500

Table 3.6 (v). Profit

	£
Total gross margin	28,100
less total fixed costs	22,500
PROFIT	5,600

Note To be precise one would deduct the rental value of the house and the use of the car from the fixed costs. However, as these estimates are so subjective it is considered more practical to show these as a separate item in the gross output.

The next step in the analysis is to calculate the gross margins achieved from the individual enterprise, in this case the dairy cows and the dairy replacements.

To do this one needs to know the number and value of the dairy heifers transferred into the dairy herd and the number and value of the calves transferred from the dairy herd to the dairy replace-

Table 3.7. Calculation of individual enterprise gross margins

	Dairy cows £	Dairy replacements £	Hay and silage and notional income £	Total £
Gross ouput				
Closing valuation	25,600	10,500	1,200	37,300
Notional income	300	—	1,300	1,600
Sales: milk	40,300	—	—	40,300
cull cows	7,400	—	—	7,400
calves	1,600	—	—	1,600
Transfers out	1,100†	11,000*	—	12,100
Sub-total (A)	76,300	21,500	2,500	100,300
Opening valuation	24,000	10,000	1,000	35,000
Purchases	1,000	—	—	1,000
Transfers in	11,000*	1,100†	—	12,100
Sub-total (B)	36,000	11,100	1,000	48,100
GROSS OUTPUT (A−B)	40,300	10,400	1,500	52,200
Variable costs				
Feedingstuffs	12,400	3,000	—	15,400
Livestock sundries	2,400	400	—	2,800
Seeds, fertilisers and other crop expenses	4,294	1,606	—	5,900
Total variable costs	19,094	5,006	—	24,100
GROSS MARGIN	21,206	5,394	1,500	28,100

* Heifers transferred into Dairy Herd, 20 at £550** per head.
† Calves transferred from Dairy Herd, 22 at £50** per head.
** *Note:* They are transferred at their estimated sale value, not at their estimated cost of production.

ment enterprise. In addition one has also to be able to allocate the variable costs, *i.e.*, feedingstuffs, livestock sundries, etc., between these two enterprises.

Finally one also needs to know the average number of stock carried throughout the year so that the results can be expressed on per head basis. Given this information the gross margin results for this example farm can be analysed as shown in Table 3.7.

The allocation of the feed and livestock sundries to the dairy cows on the one hand and the replacements on the other is relatively simple *providing* that the necessary records are kept during the year. The allocation of the seeds, fertiliser and other crop variable costs is much more difficult. This is usually resolved by making an allocation on a livestock unit basis as illustrated in Table 3.8. There are inherent weaknesses in the method of allocation and this is discussed in more detail later, see Chapter 8.

Table 3.8. Allocation of crop variable costs

Livestock Units	no.	Livestock unit factor	No. units
Heifers over 2 years	8	0·8	6·4
Heifers 1–2 years	16	0·6	9·6
Heifers under 1 year	18	0·4	7·2
			23·2
Dairy cows	62	1·0	62·0
TOTAL	—	—	85·2

Forage costs (40 hectare)	£
Seeds	400
Fertilisers	5,000
Sprays and sundries	500
TOTAL	5,900

Forage costs per livestock unit	£5.900 ÷ 85·2 = £69·24
No. livestock units per hectare	85·2 ÷ 40 = 2·13

Allocation	£		Hectares
To dairy cows	£69·24 × 62 = 4,294		29·1 (62 ÷ 2·13)
To dairy heifers	£69·24 × 23·2 = 1,606		10·9 (23·2 ÷ 2·13)
		5,900	40·0

PROVIDING THE INFORMATION

To produce accounts in the form described and as shown in Appendices 1 and 3, it is necessary to have a good book-keeping system but the time that needs to be spent on book-keeping need not be substantial if properly organised. To produce accounts in the form described for a farm of 80–200 hectares, a well trained secretary should only need to make twelve half-day, *i.e.*, monthly visits to the farm.

A *Cash Analysis Book* is essential and the headings used in this book should conform with those shown in the trading account earlier in this chapter. A farmer setting up such a system for the first time would be advised to employ a secretary and/or to purchase the NFU/ADAS Record Book which has been purposely designed to record the information required for farm business analysis.

To provide the information required for gross margin analysis there is a need for physical as well as financial information and records have to be kept so that variable cost items can be allocated to individual enterprises.

The biggest problem is usually to do with the allocation of concentrate feeds. If there is only a dairy herd and a replacement enterprise and all food is purchased there is little or no difficulty. The problem increases when there are other livestock enterprises, home-grown grain is fed and rations are made up on the farm. In this case some form of *Feed Recording System* is essential. The system needs to be designed to meet the needs of the individual farm and must be designed so that the records of what is said to have been fed can be reconciled to actual deliveries and stocks on hand at the beginning and end of the month.

This feed record should also be used to compare what has actually been fed to any group of stock to what it was intended should have been fed.

COMPARATIVE ANALYSIS

Having analysed and processed our accounts into gross margins and fixed costs we are now in a position to compare our results to those achieved by other farms costed on the same basis. *Note*, we can also compare the results to those we hoped to achieve and this forms the basis of the budgetary control system described later in the book in chapters 10 and 11.

When comparing our results to standards we need to be aware of the limitations of the standard data and the need to try to ensure that like is being compared to like. Standard data for the whole farm is normally presented on a per hectare basis. The gross margins, fixed costs and profitability levels per hectare tend to be higher the smaller the size of the farm. It is important, therefore, to try to compare the results for a particular farm to those of other farms of a similar size.

In periods of inflation and rapidly changing levels of profitability from year to year it is also important to compare results to those achieved by other farms in the *same* year. This can pose a problem as the comparative data is not usually available until some 9–15 months after the end of an accounting year.

It also needs to be borne in mind that the standard data covers a very wide range of performance. The results in most surveys are therefore divided to show the results for the top 25–30 per cent of farms as well as the results for the average, see for example the Manchester University data quoted earlier in this chapter.

Bearing in mind these limitations much can be gained from a comparison of the results from an individual farm to standards. Much can also be gained by studying the differences between the most profitable and average farms within a group. Note for example that the most profitable farms in the Manchester University survey achieve much higher gross margins but incur very little more expenditure on fixed costs than the average farm.

When comparing the results for an individual farm to standards the first step is to look at the results for the farm as a whole to see whether the individual farm has a higher or lower Net Farm Income or Management Investment Income than that achieved by the comparative farms. The next step is to continue the comparison to see whether this advantage or disadvantage is due to differences in the fixed costs structure or to differences in the gross margin. One of the major objectives in the comparison is to determine where scope lies to increase profits. If fixed costs are higher than average this will suggest that these should be examined to see if they can be reduced. On most farms, however, the main scope for improving profits is by increasing the gross margin and this aspect is discussed in more detail in the next chapter.

FIXED AND VARIABLE COSTS

Before moving on to a detailed discussion of profitability factors an explanation is necessary of the concept of fixed and variable costs (for formal definitions see Appendix 2).

The items which are usually treated as fixed costs are detailed earlier in this chapter. Fixed costs are those costs which do not change as a result of a small change in the organisation of a business, and they are also difficult to allocate to an individual enterprise from a costings point of view. If for example a farmer decides to keep 65 cows instead of 62 it is unlikely that this would have any direct effect on the cost of labour, rent, rates, power and machinery costs or on general overheads. It would, however, immediately result in a need for more concentrate feeds and sundry costs such as A.I. fees. Whether it would immediately lead to an increase in seed and fertiliser cost is detatable, depending on whether or not there was a small surplus of grass and conserved foods available at the present level of inputs, but in the long term the cost of these items would tend to rise in proportion to the increase in cow numbers.

The significance of the above is that the effect on the profit of the additional three cows can be assessed *without* needing to know what the fixed costs are per cow. The increase in the profit due to keeping the additional cows will be equal to the gross margin that can be expected from these three cows. Keeping an extra three cows will lead to a *reduction in the fixed costs per cow but the total fixed costs will remain the same.*

There is one exception to this rule, however, and that is the effect of the increase in cow numbers on finance or interest charges. Interest charges are usually treated as part of fixed costs because they cannot be readily allocated to an individual enterprise but in a borrowed capital situation a small increase in cow numbers will immediately lead to an increase in interest charges. The increase in profit from the additional cows is therefore equal to the gross margin they produce less the interest charges on the capital borrowed to purchase the cows.

It is important to remember that the effect on the profit of a change in cow numbers is not so easily assessed if the change in numbers is more substantial. An increase in cow numbers from 62 to say 80 would probably also necessitate changes in building equipment and may lead to the need for more overtime hours. Certain overhead costs would, however, stay the same and the increase in fixed costs *would not be proportionate to the increase in cow numbers.*

This concept of fixed and variable costs is fundamental to the understanding of farm business management principles, costings and budgeting methods and is discussed again later in the book. It also explains why very few complete costings are carried out in farm business management. Knowing how much 'profit' is made from an individual dairy cow is of little value as the change in profit of the whole farm as a result of changing dairy cow numbers is *not* proportionate to the profit made per cow. This is the basic reason for the development of the gross margin costing system which has the added advantage of being much less time-consuming than a full enterprise costing system.

THE FARM BALANCE SHEET

The farm balance sheet performs two functions. Firstly, as its name implies, it provides a means whereby the books or accounts for a particular year can be balanced. The objective in this instance is to reconcile the profit shown in the trading account with the balance or overdraft at the bank and with the change that has taken place in other assets and liabilities. For the farm described earlier in this chapter this part of the balance sheet statement could appear as shown in Table 3.9.

Table 3.9. Change in net capital (worth) over a year

	£
Net capital (worth) at start of year	40,000
add	
Profit for year	5,600
Private capital introduced	600
	46,200
deduct	
Private drawings during year	6,200
Tax payments	1,200
	7,400
NET CAPITAL (WORTH) AT END OF YEAR	38,800

This statement shows a reduction in the Net Capital or Net Worth of the business because the private drawings plus tax payments exceed the profit made during the year.

The second function of the balance sheet is to show in some detail the assets and liabilities of the business and how the net worth is determined.

When presenting a balance sheet it is advantageous to have the assets listed in a descending order of liquidity, that is the ease with which the capital invested in them can be realised. The liabilities should also be listed in ascending order in which they are due for payment, that is with trade creditors and the bank overdraft at the top and long-term loans at the bottom.

It is also advantageous to show the balance sheets for several years alongside each other so that the trend in net worth and in the ratio of assets to liabilities can be readily identified. A form that can be used for this purpose is shown as Table 3.10.

The arbitrary figures shown in the table for 1978 and 1979 are for a 80 hectare tenanted farm. Between 1978 and 1979 there is an increase in total assets from £98,000 to £110,000. Short-term liabilities have increased by £5,000 and there are no long-term liabilities, so the net worth is up by £7,000. In both 1978 and 1979 the business is in a strong liquidity position because current assets are well in excess of short-term liabilities. There is, for example, £8,000 saleable crop produce on hand at the 31 March 1979 and the farmer could use this if he wished to repay most of the bank overdraft.

1979–80 is assumed to be another profitable year. The net worth goes up by a further £8,000 and the farmer purchases 20 hectares of adjoining land for £100,000. Of this, £5,000 is funded from profits, £80,000 by an AMC loan and £15,000 by an increase in the overdraft. At this stage the liquidity position still seems reasonable as the current assets still exceed the short-term loans. By 1981, however, a different picture has emerged. Expenditure has been incurred on buildings and equipment, and on breeding livestock, and the total fixed assets have gone up by a further £19,000. An attempt to fund some of this out of profits has been unsuccessful, and the liquidity position has worsened considerably. There is no saleable crop produce on hand, consumable stores are down as funds have not been available to purchase fertilisers in advance as was the previous practice. Current assets are now £9,000 less than short-term loan liabilities. This business is now in serious trouble as too great a proportion of the investments is in fixed assets. The position may be eased by replacing some of the short-term loans by long-term loans but the basic problem is that the service charges, that is the interest charges on the borrowed capital, are more than the business can afford. To avoid getting into this situation careful budgeting and planning is essential.

How this can be done is shown in Chapter 11.

Table 3.10. Statement of assets and liabilities

	80 hectares: all tenanted		80 hectares as tenant + 20 hectares owned	
	31.3.78 £	31.3.79 £	31.3.80 £	31.3.81 £
ASSETS				
Current				
Cash in hand and bank	—	—	—	—
Debtors	8,000	8,000	8,000	10,000
Saleable crop produce	7,000	8,000	8,000	Nil
Consumable stores	6,000	8,000	8,000	4,000
Growing crops and tillages	5,000	6,000	6,000	8,000
Trading livestock	18,000	20,000	20,000	16,000
Total current assets	44,000	50,000	50,000	38,000
Fixed				
Breeding livestock	36,000	40,000	45,000	50,000
Machinery and equipment	18,000	20,000	20,000	24,000
Buildings and fixtures	—	—	—	10,000
Land	—	—	100,000	100,000
Total fixed	54,000	60,000	165,000	184,000
TOTAL ASSETS	98,000	110,000	215,000	222,000
LIABILITIES				
Short-Term Loans				
Creditors	8,000	8,000	10,000	12,000
Bank overdraft	5,000	10,000	25,000	25,000
Hire purchase	—	—	—	10,000
Other short-term loans	—	—	—	—
Total short-term	13,000	18,000	35,000	47,000
Long-Term Loans				
Bank loans	—	—	—	—
AMC loans	—	—	80,000	80,000
Total long-term	—	—	80,000	100,000
Total loans	13,000	18,000	115,000	127,000
NET WORTH	85,000	92,000	100,000	95,000
TOTAL LIABILITIES	98,000	110,000	215,000	222,000

Chapter 4

WHOLE FARM AND FIXED COST PROFITABILITY FACTORS

SYSTEM EFFICIENCY AND ENTERPRISE MANAGEMENT EFFICIENCY

The previous chapter has shown how the profit made by a farm equals the excess of gross margin over fixed costs. Chapter One described the functions involved in management, indicating that a manager has basically two jobs to do: firstly, to decide on the farm policy and secondly to put this into operation. It follows that the profit made on a farm also depends on how well he does these two jobs.

The best profits are made when there is a good farming system and this is effectively managed.

The relationship between farming profitability, gross margins, fixed costs and the concept of system efficiency and enterprise management efficiency is illustrated in Fig. 3. This diagram shows how the various husbandry, economic, management and other factors determine the profit of a dairy farm. The successful dairy farmer understands these rather complex interrelationships and is able to bring them together to make a success of the dairy farm business as a whole.

The effect on the farm profit of differences in system efficiency and enterprise management efficiency is illustrated in relation to a part-owned, part-tenanted seventy-hectare farm by the data set out in Table 4.1.

Farmer A and Farmer B have the same farming system, *i.e.*, same number of cows and young stock but the profit made by Farmer B is only 5·8 per cent of that made by Farmer A. This 94·2 per cent reduction in profit is the consequence of a 10 per cent reduction in gross margins and 10 per cent increase in fixed costs in turn due to an assumed lower level of enterprise management capability.

Farmer A and Farmer C have the same level of enterprise efficiency (*i.e.*, gross margins per cow and per head of young stock)

50

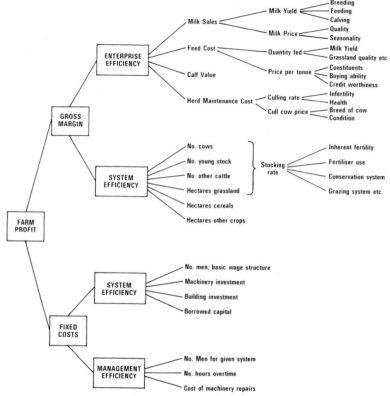

Fig. 3. Factors determining farm profits

and the same fixed cost *but* the profit made by Farmer C is only 19 per cent of that made by Farmer A. This 81 per cent reduction in profits simply reflects an assumed 10 per cent reduction in stock numbers, *i.e.*, a 10 per cent less efficient farming system.

'RENT EQUIVALENT'

The term 'Profit Before Rent and Finance Charges' has purposely been used in the financial illustration given in Table 4.1 so as to allow the introduction of the term **Rent Equivalent**. One of the limitations of most comparative accounts data is the treatment of all farms as 'tenants' even though 70 per cent of land is now owner-occupied. Net Farm Income and Management and Investment Income are useful, particularly from an academic point of

Table 4.1. The effects of different system and enterprise management efficiencies

	A Good system and good enterprise management			B Good system but poor enterprise management			C Poor system but good enterprise management		
	no.	£ per head	£	no.	£ per head	£	no.	£ per head	£
Gross Margin									
Dairy cows	100	400	40,000	100	360	36,000	90	400	36,000
Young stock	76	100	7,600	76	90	6,840	68	100	6,800
		£ per hectare			£ per hectare			£ per hectare	
		680	47,600		612	42,840		583	40,800
Fixed Costs									
Labour inc. farmer		200	14,000		220	15,400		200	14,000
Power & machinery		200	14,000		220	15,400		200	14,000
Sundry overheads		50	3,500		55	3,850		50	3,500
		450	31,500		495	34,650		450	31,500
Profit Before Rent and Finance Charges		230	16,100		117	8,190		133	9,300
Rent and rates		40	2,800		40	2,800		40	2,800
Finance charges		70	4,900		70	4,900		70	4,900
'Rent equivalent'		110	7,700		110	7,700		110	7,700
PROFIT MARGIN		120	8,400		7	490		23	1,600

view, in the assessment of management efficiency but on the farm the main criterion is the profit remaining after meeting the 'Actual Rental Equivalent'.

The 'actual rental equivalent' (*i.e.*, rent actually paid plus finance charges), does not often equate to the 'True Rental Value' of the holding. It is more likely to reflect the length of time the tenant has been on the holding and how recently he has purchased additional land or carried out improvements. Very often one finds that farms with poor buildings and equipment have high actual rent and finance charges, whereas farms with completely modern set-ups have little or no rent or finance charges to pay. The current (1981) value of the buildings and fixed equipment on a well-equipped dairy farm is likely to be in the region of £800–£1,000 per cow but may only have cost £200–£250 per cow. If a 'rental value' is charged based on the value of these buildings as well as the land then the resulting figure is likely to be in excess of £100 per cow or more than £250 per hectare compared to actual rents in the region of £75–£125 per hectare.

It is usually assumed that the 'rent not paid' for owner-occupied land can be used to service finance charges on borrowed capital. This is often an erroneous assumption because the 'rent' is frequently required to service property repairs and ownership expenses. In particular it is required to service improvements.

This point needs to be carefully borne in mind by anyone contemplating starting farming or acquiring additional land.

LABOUR COSTS

The Agricultural Wages Board determines annually the wages to be paid to agricultural workers. Statutory rates are determined for ordinary workers, craftsmen, grade II workers and grade I workers. Prior to 1980 the rates determined for craftsmen grade II and grade I workers were 110, 120 and 130 per cent respectively of the

Table 4.2. Labour costs

Craftsmen's basic wage per week		Milk price			Milk price to wages ratio†	Ratio 1979 =100
	£	Year ended	Per litre p	Per 1000 litres £		
Jan. 1975	31.35	March 1976	7.82	78.20	2.49	116
,, '76	40.15	,, '77	9.27	92.70	2.31	108
,, '77	42.90	,, '78	9.83	98.30	2.29	107
,, '78	47.30	,, '79	10.67	106.70	2.26	106
,, '79	53.35	,, '80	11.44	114.40	2.14	100
,, '80	66.70	,, '81	12.50*	125.00	1.87*	87
,, '81	73.60					

* Estimate.
† Value of 1,000 litres milk divided by one week's basic wage.

ordinary worker rate. In 1980 these were increased to 115, 125 and 135 per cent respectively.

Table 4.2 shows the wage rates for craftsmen as determined by the Agricultural Wages Board alongside the wholesale milk price as received by the farmer. It demonstrates quite forcibly one of the major problems facing dairy farmers in the 1980s, namely the substantial increase in wage rates relative to milk prices.

In Table 4.3 the annual costs of employing a herdsman plus his relief milker in 1980 is shown alongside that incurred in 1978. A 66-hour week is assumed, *i.e.*, ten hours per day during the week and eight hours per day at weekends.

Table 4.3. Herdsman's costs 1978 and 1980

	1980	*1978*
Craftsman's wage for 40 hours	66.70	47.30
26 hour's overtime at £2.50	65.00	46.12
	131.70	93.42
	× 52 weeks	× 52 weeks
	6,848	4,858
Add 14 per cent employer's national insurance contribution	959	680
	7,807	5,538

In 1978 a herd of eighty cows with an average yield of 5,000 litres would have produced a total milk income of £42,680 (80 × 5,000 litres × 10.67 pence per litre). The annual cost of the herdsman including his relief milker represents 13 per cent of this output.

The milk income needed in 1980 to achieve a labour cost equal to 13 per cent output is £60,054. At 12.50 pence per litre this is equivalent to 480,432 litres, equivalent in turn to eighty cows at 6,005 per cow, or ninety-six cows at 5,000 litres.

The message is quite clear: a 20 per cent improvement in productivity is needed to maintain milk sales/labour cost ratios at their previous levels. Whether this is achieved by more cows, by an increase in yield per cow or by both together will depend on the individual farm circumstances. If it is not achieved on your farm then one more farm will eventually be added to the list of those going out of milk production.

POWER AND MACHINERY COSTS

A reminder is needed first about machinery and equipment depreciation and the points discussed in Chapter 3, page 35 in relation to inflation. In most farm accounts depreciation is determined on the historic costs basis, and consequently the depreciation charges shown need to be at least doubled to arrive at a realistic figure.

Power and machinery costs, like labour costs, have risen at a more rapid rate than milk prices and control of these costs is another major challenge to dairy farmers. A feature of the development of dairy farming in recent years has been the increasing sophistication and size of tractors and forage harvesting equipment. Consequently, dairy farms often have very high power and machinery costs as well as high labour costs relative to other farming systems. This is illustrated in Table 4.4 by data taken from the 'Report on Farming in the Eastern Counties of England 1979–80,' produced by the University of Cambridge.

Table 4.4. Labour and machinery costs on arable and dairy farms (eastern counties)

Farming type	Mainly cereals £ per hectare	Mixed cropping £ per hectare	Arable with dairy £ per hectare
Paid labour	64.2	103.3	111.5
Farmer's manual labour	9.8	6.7	19.2
Total labour	74.0	110.0	130.7
Power and machinery	128.4	165.0	142.5
Labour, power & machinery	2,020.4	275.0	273.5
Gross margin	377.3	521.4	419.4
	£	£	£
Gross margin per £100 labour	510	474	320
Gross margin per £100 labour & machinery	186	190	153
	percentage of area	percentage of area	percentage of area
Cereals	73.1	60.7	38.6

The mainly cereal farms have a labour cost of only £74.0 per hectare compared to £110.0 per hectare on the more intensive mixed cropping farms and £130.7 on the arable-with-dairy farms. The gross margin per £100 labour is £510 on the mainly cereal farms and £474 on the mixed cropping farms, but on the dairy-with-arable farms it is only £320.

Similarly the gross margin per £100 labour and machinery is £186 on the mainly cereal farms and £190 on the mixed cropping, but only £153 on the arable-with-dairy farms.

These figures help to explain why there has been a gradual run-down in the number of dairy herds in the eastern counties. On farms suited to arable production higher returns can be generated relative to labour and machinery inputs from arable crops than can be made from dairy farming.

The figures quoted are for one year only, and for a difficult dairy farm year, but similar trends have been evident in previous years. The result is that dairy farming is becoming concentrated in the western half of the country on land suited to grassland production but not to arable farming.

PRODUCTIVITY PER £100 LABOUR AND MACHINERY COSTS

The gross margins achieved per £100 labour and per £100 labour and machinery costs are considered to be good means of assessing the productivity of labour and machinery. The results achieved by dairy farms costed by Manchester University in 1977, 1978–9 and 1979–80 are shown in the Table 4.5. These farms are in the counties of Lancashire, Cheshire, Shropshire and Staffordshire, and the costing period includes the very severe winter of 1978–9.

The results shown in this table highlight the care that needs to be taken when trying to compare the productivity of one farm with another. It is important, for example, to determine whether or not the gross margin figure does or does not include 'breeding livestock appreciation'.

It is also essential to realise how difficult it is to give a rule-of-thumb guide to productivity as the results vary enormously from year to year due to climate and economic factors. Mention has already been made in this chapter of the substantial rise that has taken place in labour and machinery costs relative to milk prices. This, coupled with the severe winter of 1978–9, led to a reduction in the gross margin per £100 labour achieved by these farms from £468 in 1977–8 to only £341 in 1979–80.

As with all aspects of management one can expect considerable differences between farms in terms of labour and machinery pro-

Table 4.5 Actual productivity comparisons

	1977–8 £ per hectare	1978–9 £ per hectare	1979–80 £ per hectare
Total gross margin	542	626	566
Breeding livestock appreciation	65	97	34
Gross margin excluding BLA	477	529	532
Paid labour	80	94	117
Farmer's manual labour	36	43	49
Total labour	116	137	166
Power and machinery	120	152	187
Labour and machinery costs	236	289	353
	£	£	£
Gross margin per £100 labour	468	458	341
Gross margin per £100 labour and machinery	230	216	160
Gross margin (ex. BLA) per £100 labour	412	386	320
Gross margin (ex. BLA) per £100 labour and machinery	202	183	150

ductivity. This is confirmed by the data given in Table 4.6 in respect of the over-fifty-hectare farms costed by Manchestery University.

This data also highlights other differences that are found between the most profitable and the average farm.

The more profitable farms keep more cows and produce more milk per cow than the average farm. The stocking rate is much higher on the high-profit farm than it is on the average farm although young stock numbers are no greater.

The total fixed inputs at £466 per hectare are £74 more than those on the average farm but this is more than offset by an increase in the gross margin of £258 per hectare, giving an increase in the Management and Investment Income of £164 per hectare.

Attention is drawn to the significance of Breeding Livestock Appreciation in the results for this particular year when cow values rose by about £70 per head. At £97 per hectare it represents 42 per cent of the Management Investment Income achieved by the average farm. On the most profitable farms the Breeding Livestock Appreciation is £129 per hectare and represents 32 per cent of

Table 4.6. Productivity in 50 hectare-plus dairy farms

| | 1978–9 | |
	Average	Above Average
Number of farms	38	6
	£ per hectare	*£ per hectare*
Gross margin (including BLA)	626	864
Gross margin (excluding BLA)	529	735
Total labour	137	168
Power and machinery	152	177
Labour and machinery	289	345
Gross margin (ex BLA) per £100 labour	*£* 386	*£* 437
Gross margin (ex BLA) per £100 labour and machinery	183	213
Rent and rates	66	80
Property repairs and maintenance	17	22
General overhead costs	20	19
Total fixed inputs	392	466
Management and investment income	234	398
	No.	*No.*
Farm size (*hectares*)	82.1	80.5
Cows per 100 hectares	134	177
Other cattle per 100 hectares	57	54
Stocking rate (*livestock units per hectare*)	2.27	2.70
Valuation per Hectare	*£*	*£*
Crop produce	25	19
Livestock	679	832
Machinery	307	346
Stocks feed, fertilisers etc.	57	82
	1,068	1,279
Milk yield (*litres per cow*)	5,262	5,984
Milk sales per cow (*£*)	558	642

their Management and Investment Income. Once again the reader is reminded that the value of cows as a hedge against inflation is a point that needs to be kept in mind when comparing the profitability of dairy cows to other enterprises, particularly to arable crops.

VARIATIONS IN PROFITABILITY BETWEEN FARMS

The enormous variation found in the financial results between farms is also illustrated by data taken from the Milk Marketing Board's Report No. 24. *An analysis of F.M.S. costed farms in 1979–80.* The range in results between the top 25 per cent and botton 25 per cent selected on the basis of profit per hectare is shown in Table 4.7.

Table 4.7. The variation in profitability between farms, 1979–80

	Top 25 per cent £		Bottom 25 per cent £	
	£ total	per hectare	£ total	per hectare
Gross Margins				
Dairy cows	38,543	488	26,977	303
Other grazing livestock	6,417	81	1,918	21
Other enterprises	1,034	13	2,397	27
Other farm income	983	12	1,351	15
TOTAL	46,977	594	32,643	366
Fixed Costs				
Paid labour (only)	6,752	85	10,173	114
Power and machinery: cost	6,575	83	8,162	91
Power and machinery: depreciation	4,109	52	4,584	51
Property charges: cost	4,553	58	4,640	52
Property charges: depreciation	2,698	34	3,094	35
Sundry overheads	3,405	43	3,992	45
Total before finance	28,092	355	34,645	389
Profit before finance charges	18,885	239	(2,002)	(23)
Interest charges	1,679	21	8,565	96
PROFIT (LOSS), BEFORE FARMER'S WAGES	17,206	218	(10,567)	(119)
Average size	79 hectares		89 hectares	

The least profitable farms have higher labour and machinery costs per hectare even though they are slightly larger than the more profitable farms.

The gross margin per hectare achieved by the less efficient farms is only 61 per cent of that achieved on the most efficient farms. This low gross margin reflects weaknesses in both enterprise efficiency and system efficiency. The former is shown up by the low gross margin per cow and the low gross margin per other grazing livestock unit (see Table 4.8). The low number of cows and live-stock units per hectare shows up the weakness in system efficiency.

Table 4.8. Gross margins per cow and other livestock units, 1979–80

| | Top 25 per cent | | | Bottom 25 per cent | | |
	no.	margin per cow £	margin per unit £	no.	margin per cow £	margin per unit £
Gross Margin						
Dairy cows	115	335	38,543	107	252	26,977
Other grazing livestock units	46	139	6,417	43	45	1,918
			45,060			28,895
Total livestock units	161			150		
No. hectares	79			89		
Livestock units per hectare	2.04			1.68		

The actual reasons for these differences in performance cannot be ascertained from survey data without investigating individual farm circumstances. It would appear that lack of capital was a significant factor in this instance as the finance charges on the bottom 25 per cent are more than four times those incurred by the top 25 per cent.

This tentative conclusion is supported by an analysis of the balance sheets of the bottom and top farms shown in Table 4.9. The 'operating capital' on the bottom farms of £95 per hectare is only £136 more than the total liabilities (borrowing). The operating capital invested on the top farms is £270 more than on the bottom farms and is £964 per hectare more than their liabilities.

It was not intended to discuss capital so early in this book; capital is covered in more detail in Chapter 13, but having introduced this subject we need to continue.

The bottom 25 per cent of farms are faced with a dilemma. They are producing a poor result partly because they have too little investment in livestock but they will need to borrow even more capital if they are to bring their stocking rate up to that of the top 25 per cent. To do this they may also need to incur capital expenditure on landlord's or tenant's fixtures. This lack of investment may explain in part why they have high labour costs although this is not borne out by the depreciation charges. These are very similar to those incurred by the top 25 per cent.

Table 4.9. Capital structures: top and bottom 25 per cent, dairy farms, 1979–80

	Top 25 per cent		Bottom 25 per cent	
	£	£ per hectare	£	£ per hectare
Assets				
Debtors	8,583	108	7,146	80
Crops, stores and tillages	8,845	112	7,072	80
Livestock	60,616	767	49,683	558
Tenant's fixtures and machinery	18,534	235	20,843	234
Total operating capital	96,578	1,222	84,744	952
Landlords' fixtures	9,082	115	10,466	117
Freehold value of farm	190,757	2,415	206,329	2,318
Total capital invested	296,419	3,752	301,494	3,387
Liabilities				
Creditors	7,030	89	10,062	113
Bank overdraft	10,810	137	35,014	393
	17,840	226	45,076	506
Mortgages	327	4	19,282	216
Hire purchase	294	4	1,385	16
Other liabilities	1,912	24	6,898	78
Total liabilities	20,372	258	72,641	816
NET WORTH	276,046	3,494	228,853	2,571

The gross margin results achieved per cow by these farms are shown in Table 4.10.

The results achieved by the top 25 per cent are consistently better than those achieved by the bottom 25 per cent so one begins to question whether this is the result of the shortage of capital or whether the capital shortage is a result of poor performance.

These results bear out those found in many surveys, *i.e.*, the ability of the best farmers to achieve considerable increases in output with minimal increases in inputs compared to their less successful neighbours. They achieve this in part due to having more capital resources and better farms at their disposal but in most instances the major part is explained by better management.

The best farm, for example, in this case achieves £80 more milk sales for an additional expenditure on feed of £15, giving a margin over feed costs that is £65 more than that found on the worst farm.

This additional margin of £65 is achieved with only £1 more expenditure on forage per cow and at a stocking rate of 2·08 livestock units per hectare instead of 1·85.

Table 4.10. Gross margins per cow: top and bottom 25 per cent dairy farms 1979–80

	Top 25 per cent £	Bottom 25 per cent £
Milk sales	658	578
Calf output	57	45
	715	623
Herd depreciation	33	34
Gross output	682	589
Variable Costs		
Concentrates	232	227
Purchased bulk feeds	33	23
Sundries	29	35
	294	285
Gross margin before forage costs	388	304
Forage costs	54	53
GROSS MARGIN	334	251
No. cows per hectare	2.08	1.85
GROSS MARGIN PER HECTARE	695	464
Milk yield per cow	5,739 litres	5,066 litres
Milk sale price per litre	11.47p	11.41p
Concentrates and purchased feed cost per litre	4.62p	4.93p
Margin over concentrates and purchased feed per litre	6.85p	6.48p
Margin over concentrates and purchased feed per cow	£393	£328
Cost of concentrates per ton	£119	£121
Sale price per calf	£60	£53
Sale price per cull cow	£299	£277
Herd replacement rate	22·6 per cent	25·2 per cent
Nitrogen use per hectare	300 kg	195 kg

VARIATIONS IN PROFITABILITY FROM YEAR TO YEAR

The variations in the profitability of dairy farming have been very marked in recent years. These variations are largely due to seasonal and economic factors over which the individual farmer or manager has little or no control, but nonetheless he needs to be aware of them.

Mention has already been made in this chapter of the substantial change that took place in 1980 in the ratio between the price of milk and the cost of fixed inputs, particularly labour. Published results for the year ending 31 March 1980 show substantial falls in profitability compared to the previous year. Relatively low levels of profitability can also be expected to be shown when results are available for 1980–1.

It needs to be recognised, however, that these poor years will almost certainly be followed in due course by good years. This cyclical nature of farming profitability is a factor with which the dairy farmer has to come to terms.

The profitability of dairy farming is very closely linked to the profitability of beef production due to its effect on calf, and perhaps more important, cull cow prices. The latter is particularly important as it has such an important effect on the asset values of a dairy farm, *i.e.*, the livestock valuation. This in turn largely determines the creditworthiness of a dairy farmer's business, particularly in the case of a tenant farmer. Dairy farmers can expect to make good profits when beef prices are rising, as was the case in 1971–2, 1972–3 and 1978–9. Conversely they make below-average profits when beef prices are depressed, as for example in 1974–5 and 1979–80.

Changes in the relationship between the price of milk and the cost of concentrates are also responsible for considerable fluctuations in the profitability of dairy farming.

When dairy farming is relatively profitable one can expect the price of milk per litre to be nearly the same as the cost of concentrates per kilogram. For example, given a milk price of 13.5 pence per litre, one would hope that concentrate feed costs could not be more than £140 per tonne or 14.0 pence per kg.

The relationship between milk price and feed costs in recent years is given in Table 4.11 (source *Dairy Facts and Figures*).

Data is not yet available (June 1981) for 1980–1, but when available it is expected to show a favourable milk price : feed cost ratio. Beef prices at the present time are also firm and rising. This

coupled with the improved milk price:feed cost ratio gives hope
that dairy farmers' financial results will show a substantial up turn
in 1981–2 and 1982–3.

Table 4.11. Milk price: feed costs ratios 1974–80

	Milk price (pence per litre)	Concentrate costs (pence per kg)	Ratio
1974–5	6.245	7.82	1:1.25
1975–6	7.816	8.02	1:1.03
1976–7	9.269	10.03	1:1.08
1977–8	9.827	11.06	1:1.13
1978–9	10.430	11.26	1:1.08
1979–80	11.407	12.91	1:1.13

Chapter 5

DAIRY HERD ENTERPRISE EFFICIENCY AND PROFITABILITY FACTORS

GROSS MARGIN PER COW AND PER HECTARE

The previous chapter has shown how the profitability of a dairy farm as a whole depends on the relationship between fixed costs and the total gross margin. It has also shown how the latter depends on the efficiency with which the farm is organised and managed, and has demonstrated the ability of the best farmers to produce much higher gross margins than others from a given level of fixed costs.

This chapter is concerned with the factors determing the gross margin produced by a dairy herd and with methods that can be adopted to identify weaknesses in its management.

The gross margin achieved *per cow* is accepted by many people as being the best measure of dairy cow enterprise efficiency. Grassland enthusiasts on the other hand only tend to see cows as converters of grass to £p and prefer to see gross margin *per hectare* treated as the most important measure. Gross margin per cow, however, is a much better measure of return on capital.

The various factors that influence the gross margin per cow are illustrated by budget data taken from the *Farm Management Pocket Book* by John Nix, Eleventh Edition (1981) published September 1980 (see Table 5.1).

It needs to be noted that the information shown in the table is budget data. It is based on actual results achieved for the year ended March 1980 updated to the levels expected in 1980–1.

The first thing one notices is that the budget data is shown at four differing yield levels. In the *Pocketbook* it is also shown at four differing stocking rates but only two of these are shown in our table. The data is shown in this way to demonstrate the enormous effect these two factors have on the gross margin per cow and per hectare.

The budget margin over concentrates per cow varies from £352

Table 5.1. Dairy cow gross margin standards 1981 (Source John Nix)

Performance level (yield)	Low	Average	High	Very high
Yield (*litres*)	4,250	5,000	5,750	6,250
Milk price (*p. per litre*)	12.75	12.75	12.75	12.75
Milk sales per cow (£)	542	683	733	797
Concentrates fed per litre (*kg*)	0.32	0.34	0.36	0.38
Concentrates fed (*tonne*)	1.36	1.70	2.07	2.375
Concentrate cost per tonne (£)	140	140	140	140
Concentrate cost per cow (£)	190	238	290	332
	£	£	£	£
Margin over concentrates per cow	352	400	443	465
add calf output	75	75	75	75
	427	475	518	540
subtract				
Herd depreciation	38	38	38	38
Bedding	7	7	7	7
Vet. and medicines	11	12	13	14
A.I. and recording fees	10	10	10	10
Consumable dairy stores	12	13	14	15
	78	80	82	84
GROSS MARGIN BEFORE FORAGE COSTS	349	395	436	456
Stocking rate (average)				
Forage costs per cow (£)	63	63	63	63
Gross margin per cow (£)	286	332	373	393
No. cows per hectare	1.75	1.75	1.75	1.75
GROSS MARGIN PER HECTARE (£)	502	582	654	689
Stocking Rate (High)				
Forage costs per cow	80	80	80	80
Gross margin per cow (£)	269	315	356	376
No. cows per hectare	2.50	2.50	2.50	2.50
GROSS MARGIN PER HECTARE	672	787	890	940

to £465 according to the level of milk yield. At average milk yields the gross margin varies from £502 to £689 per hectare according to stocking rate; at high yields it varies from £672 to £940 per hectare.

MILK YIELD

The budget data for 1980–1 prepared by John Nix shows margin over concentrates ranging from £352 to £465 per cow according to milk yield. That this range of performance is found in practice is supported by Table 5.2 which analyses the results according to yield of 1,104 farms costed by the Milk Marketing Board in 1979–80. This table shows a range from £207 to £429 per cow with yields ranging from 3,166 to 6,330 litres per cow.

Table 5.2. Analysis by yield of MMB costed farms 1979–80

Yield per cow (litres)	< 3,500	3,500– 3,999	4,000– 4,499	4,500– 4,999	5,000– 5,499	5,500– 5,999	> 6,000
Number of herds	24	52	116	232	314	234	132
per cent of sample	2.2	4.7	10.5	21.0	28.4	21.2	12.0
Physical results							
Herd size	71	74	90	95	102	109	108
Yield per cow (litres)	3,166	3,793	4,284	4,766	5,233	5,722	6,330
Concentrate use per cow (kg)	1,243	1,386	1,609	1,719	1,855	2,102	2,373
Concentrate use per litre (kg)	0·40	0·37	0·38	0·36	0·35	0·37	0·37
Stocking rate (LSU/ha)	1·79	1·90	1·81	1·91	1·95	2·05	2·00
Nitrogen use per ha (kg)	128	196	199	215	229	268	261
Summer milk (per cent)	56·2	55·6	53·5	51·1	49·8	49·4	48·5
Dry cow (per cent)	21·3	17·4	15·9	15·4	14·1	14·3	14·6
Replacement rate (per cent)	15·4	17·6	18·5	18·9	23·0	21·6	23·8
Financial results	£	£	£	£	£	£	£
Output per cow							
Milk sales	345.3	409·8	470·8	524·3	579·0	634·5	704·1
Calf sales	45.6	49·5	47·8	48·2	49·6	51·1	52·2
less herd depreciation	33·7	35·9	38·9	30·6	35·6	36·0	36·4
GROSS OUTPUT	357·2	423·4	479·7	541·9	593·0	649·6	719·9
Variable costs per cow							
Concentrates	137·5	155·3	177·8	191·1	208·1	237·3	274·5
Purchased bulk feed	14·1	12·2	10·8	11·5	12·3	15·8	20·6
Forage	25·9	37·3	43·1	42·4	45·5	51·2	52·2
Sundries	25·1	23·7	29·7	30·9	34·4	36·3	40·1
TOTAL VARIABLE COSTS	202·6	228·5	261·4	275·9	300·3	340·6	387·4
GROSS MARGIN PER COW (£)	154·6	194·9	218·3	266·0	292·7	309·0	332·5
GROSS MARGIN PER HA (£)	276·7	370·3	395.1	508·1	570·8	633·5	665·0
MARGIN OVER CONCENTRATE PER COW (£)	207·8	254·5	293·0	333·2	370·9	397·2	429·6
Milk price received per litre (p)	10·91	10·80	10·99	11.0	11·06	11·09	11·12
Concentrate cost per tonne (£)	113·2	112·3	111·2	111·4	112·6	113·1	116·0

F.M.S. Information Unit.
Milk Marketing Board.

Table 5.3 has been derived from Table 5.2 and shows the MoC obtained per additional litre and the average MoC per litre at the different yield levels. The MoC per litre increase from 6.56 pence at 3,166 litres to 7.09 pence at 5,233 litres, and then falls to 6.79 pence at 6,330 litres.

Although the margin per litre falls over 5,233 litres there is still an increase in the MoC of over 5 pence for every additional litre produced.

Table 5.3. Changes in margin over concentrates at different yield ranges

Yield range (litres per cow)	Yield (litres per cow)	Margin over concentrates (£)	Yield increment (litres per cow)	Margin increment (£)	Margin per 100 litres increase in yield (£)
6,000	6,330	429	608	32	5.26
5,500–5,999	5,722	397	489	26	5.32
5,000–5,499	5,233	371	467	38	8.14
4,500–4,999	4,766	333	482	40	8.30
4,000–4,499	4,284	293	491	39	7.94
3,500–3.999	3,793	254	627	46	7.34
3,500	3,166	208			

This increase in yield is also associated with an increase in stocking rate and improved gross margins per cow and per hectare.

Results are also shown in Table 5.4 for dairy farms costed by ICI in 1978–9. In this case the top 25 per cent have been selected according to their gross margin per hectare.

The top 25 per cent in this case produce 354 more litres of milk per cow than the average farmer and achieved this with no increase in concentrate costs.

Their milk sales are £48 more than the average and this accounts for virtually all the difference in the gross margin per cow.

There is no doubt, therefore, that high milk yields are very strongly associated with high margins and a high yield per cow should be a primary objective of the dairy farmer.

CONCENTRATE COSTS

This high yield has to be achieved, however, with proper control of the cost of concentrates. The main objective is to achieve a high margin not a high yield regardless of costs.

It is when one turns to concentrate costs that the problems involved in interpreting financial results begin, as will be discussed

Table 5.4. An analysis of ICI costed farms, 1978–9

	1978 Average	1979 Average	Top 25 per cent* 1979
Number of herds	125	106	26
Average number of cows	134	143	146
Average yield (*litres per cow*)	5,483	5,610	5,964
Average milk price (*p per litre*)	10.5	11.4	11.5
Concentrates fed			
(*kg per litre*)	0·30	0·32	0·29
(*tonne per cow*)	1·68	1·79	1·75
Average conc price (*£ per tonne*)	102	117	120
Silage fed (*tonne per cow*)	6·70	7·36	7·30
Hay fed (*kg per cow*)	149	105	68
	£ per cow		
Milk sales	577.5	642.6	690.4
Calf sales	49.5	51.4	52.3
Depreciation	−35.2	−36.2	−35.0
OUTPUT	591.8	657.8	707.7
Concentrates	171.5	210.2	210.4
Bulky feeds	11.4	15.3	13.5
Forage	50.3	60.5	58.4
Vet and medicines	10.3	11.4	10.6
Bedding	4.6	5.4	4.0
Other costs	18.4	21.4	24.5
VARIABLE COSTS	266.5	324.2	321.4
GROSS MARGIN PER COW	325.3	333.6	386.3
Stocking rate (*cows per hectare*)	2.09	2.07	2.36
N use (*kg per hectare*), whole farm grass	286	292	338
	£ per hectare		
GROSS MARGIN	680	691	912

* Selected on the basis of gross margin per hectare
(ICI recorded farms—1979 crop year)

more fully later. Basically what a farmer has to resolve is how far he can go in the substitution of lower-cost home-grown food for more expensive concentrates without sacrificing yield. How far he can go and the economic consequences of these decisions are central to the controversy centred round the feeding of dairy cows. The problems arise because one is trying to assess two factors at the same time, *i.e.*, the efficiency of management of the dairy cow and the efficiency of management of grassland. Differences in the efficiency with which farmers utilise grassland have an enormous effect on the profitability of dairy farms. More effective grassland management, for example, will explain in part why the top 25 per cent ICI costed farms can achieve higher yields per cow with no greater inputs of concentrates, forage or other costs per cow. These results are not exceptional; they are a common feature of surveys.

The above, however, is not all the explanation. Some farmers achieve better results than others from a given input of concentrates because these are fed at the right time. This philosophy is central to the Brinkmanship Recording and Feeding System which is described in Chapter 12. They also achieve higher yields because they have better cows and have an overall better standard of husbandry.

In recent years there has been a move towards the use of higher energy concentrates. These, of course, are more expensive but capable, if used correctly, of bringing about a higher level of production. In general, concentrates are of two types:

1. High energy concentrates—0·35 kg/litre, with an M.E. of approximately 13·5 mJ with a varying range of protein.

2. Medium energy concentrates including home-grown cereals—0·4 kg/litre with a M.E. of approximately 12·5 mJ with a varying range of protein.

There has in the past been some degree of confusion concerning the foods which come under the description of concentrates. Concentrates should include all feeds of an energy-rich nature, and certainly this would include sugar beet nuts and home-grown cereals such as barley, oats and wheat.

In some cases cereals have been used in the past in a supplementary way to provide a maintenance ration in conjunction with bulk feeds, *e.g.*, hay, kale and silage.

To have some degree of standardisation for comparative purposes, it is desirable to aggregate all concentrates in a ration and to measure their use in terms of kilogrammes per litre of milk produced.

Another area of confusion concerns the use of brewers' grains and molassed pressed pulp. These are bought-in foods; the present conventional way of assessing their use is under the heading of purchased bulk feed and their presence must always be related to the intensity of the stocking rate for the unit.

There is no allowance for purchased bulk feed in Nix's table but there is a figure in the ICI Survey.

Having reviewed the factors which make up the gross margin for the dairy herd, it is evident that there are three main areas where the farmer should concentrate and organise his management in order to improve this profit.

1. **Milk Yield**—this must be identified and a target set in the light of the farm resources and the milk production system.
2. **Feeding**—an appropriate system needs to be established which combines the level of concentrates and the quantity and quality of bulk feed to the best advantage.
3. **Stocking Rate and Fertiliser Costs**—these are closely related to the costs of feeding the cow and must be looked at in conjunction with purchased feeds and the level of output.

A number of efficiency measures and standards have been developed which are now in regular use for the purpose of controlling the progress of the dairy herd in those three main areas of management, and these are reviewed below.

MARGIN OVER CONCENTRATES (MoC)

This, the simplest of efficiency measures, is obtained by subtracting the total cost of concentrates from the total milk sales. As the milk is paid for by cheque on a monthly basis it is relatively easy to calculate the margin per calendar month (see Table 5.5).

Table 5.5. Example calculation of margin over concentrates

January 1981	Total £	No. cows	Per cow £
Milk sales	6,750	100	67.50
Concentrate cost	2,400	100	24.00
Margin over concentrate	4,350		43.50

Once twelve monthly margins have been calculated it is possible to put into operation a rolling margin over concentrates. This is simply achieved by replacing the margin of one month in the previous year and adding the margin of that month for the following year (see Table 5.6).

Table 5.6. Rolling MoC

	£ per cow
MoC for year ending September 1980	450
add MoC for October 1980	42
	492
subtract MoC for October 1979	38
	454

The total amount of concentrates should include all foods which have a high energy level and should be quite distinct from the bulk feeds used on a farm.

Home-grown cereals should all be included at an opportunity cost, *i.e.*, the price at which they could have been sold on the open market.

In the milk production areas of the west—such as Cheshire, many dairy herds receive no feeds other than concentrates, grass and conserved grass products and in these situations the progress of a dairy herd can be very effectively monitored by the margin over concentrates standard.

MARGIN OVER CONCENTRATES AND PURCHASED FEEDS

Where farms use a wider variety of foods to produce milk, the MoC figure is not comprehensive enough to be used as the sole efficiency standard. On many farms brewer's grains and molassed

Table 5.7. Margin over concentrates and purchased feeds

January 1981		Per cow £
Milk sales		55
purchased feeds:		
	concentrates	12
	bulk feed	4
		16
MARGIN		39

pressed pulp are fed. In some cases there may be a need to buy in a small amount of hay, perhaps to feed to newly calved cows to help prevent quality problems arising (see Table 5.7).

On arable farms in the eastern part of the country it is possible to find dairy farmers who use grassland for grazing during the summer, but who purchase all maintenance and production foods for the winter. In this situation the margin over purchased feeds is quite low, but the stocking density is high. A herd on this system of production could be stocked at five cows per hectare. The margin per cow might be low at £200–300 but the resultant £1,000–£1,500 margin per hectare is high. It is, however, rather pointless to compare figures achieved on a farm of this type with those found on the more typical dairy farms in the west.

AVERAGE MILK PRICE

The milk price is shown in Table 5.1 at 12.75 pence per litre, and this represents the price received *after* the MMB has deducted the transport costs, capital contribution and co-responsibility levy. If the milk price was not shown net of these items they would have to be included as part of the variable costs. It is normal practice to work to the price net of these items and this should be remembered when you are comparing your results to standards. Remember, however, to 'contra' the other items on the milk cheque, such as AI fees and treat these as variable costs.

Farmers who have recently installed a bulk tank and/or become brucellosis-accredited receive additional premiums. The brucellosis premium is equal to 0.176 p. per litre payable for five years, but this scheme became closed to new members on 31 December 1979. Premiums paid in respect of bulk tanks vary according to the size of the tank. These are paid for three years but these too ceased after the end of July 1979 when the whole country 'went bulk'.

COMPOSITIONAL QUALITY PAYMENTS

The price an individual farmer receives for his milk is determined according to the compositional quality of the milk. A new pricing scheme placing increased emphasis on milk compositional quality was introduced by the Milk Marketing Board in May 1980. The price differentials according to 'butterfat' and 'solids-not-fat' for the year 1980–1 were as shown in Table 5.8 and 5.9.

Table 5.8. Butterfat classification from 1 May 1980

Supply with the following latest average butterfat percentage	Class	Additions (+) or deductions (−) to or from the basic price (pence per litre)
For each 0·10 add one Class above Class 32	33 and higher	+ 0·135 for each Class higher than Class 40
5·50 and less than 5·60	32	+2·160
5·40 ,, 5·50	31	+2·025
5·30 ,, 5·40	30	+1·890
5·20 ,, 5·30	29	+1·755
5·10 ,, 5·20	28	+1·620
5·00 ,, 5·10	27	+1·485
4·90 ,, 5·00	26	+1·350
4·80 ,, 4·90	25	+1·215
4·70 ,, 4·80	24	+1·080
4·60 ,, 4·70	23	+0·945
4·50 ,, 4·60	22	+0·810
4·40 ,, 4·50	21	+0·675
4·30 ,, 4·40	20	+0·540
4·20 ,, 4·30	19	+0·405
4·10 ,, 4·20	18	+0·270
4·00 ,, 4·10	17	+0·135
3·90 ,, **4·00**	**16**	**NIL**
3·80 ,, 3·90	15	−0·135
3·70 ,, 3·80	14	−0·270
3·60 ,, 3·70	13	−0·405
3·50 ,, 3·60	12	−0·540
3·40 ,, 3·50	11	−0·675
3·30 ,, 3·40	10	−0·810
3·20 ,, 3·30	9	−0·945
3·10 ,, 3·20	8	−1·080
3·00 ,, 3·10	7	−1·215
2·90 ,, 3·00	6	−1·350
2·80 ,, 2·90	5	−1·485
2·70 ,, 2·80	4	−1·620
2·60 ,, 2·70	3	−1·755
2·50 ,, 2·60	2	−1·890
Less than 2·50	1	−2·025

Milk Compositional Quality is affected by the breed of cow and typical average figures are shown in Table 5.10.

The price received by most producers is close to the average but inefficient producers may find they are suffering considerable price penalties. Milk compositional quality is most affected by breed and those with Channel Island cattle receive a substantial price premium which tends to offset the lower yields from these breeds.

Table 5.9. SNF classification from 1 May 1980

Supply with the following latest average snf percentage		Class	Additions (+) or deductions (−) to or from the basic price (pence per litre)		
			Normal	Low-snf	Total
10·20 and less than	10·30	A	+1·230	NIL	+1·230
10·10 ,,	10·20	B	+1·148	NIL	+1·148
10·00 ,,	10·10	C	+1·066	NIL	+1·066
9·90 ,,	10·00	D	+0·984	NIL	+0·984
9·80 ,,	9·90	E	+0·902	NIL	+0·902
9·70 ,,	9·80	F	+0·820	NIL	+0·820
9·60 ,,	9·70	G	+0·738	NIL	+0·738
9·50 ,,	9·60	H	+0·656	NIL	+0·656
9·40 ,,	9·50	I	+0·574	NIL	+0·574
9·30 ,,	9·40	J	+0·492	NIL	+0·492
9·20 ,,	9·30	K	+0·410	NIL	+0·410
9·10 ,,	9·20	L	+0·328	NIL	+0·328
9·00 ,,	9·10	M	+0·246	NIL	+0·246
8·90 ,,	9·00	N	+0·164	NIL	+0·164
8·80 ,,	8·90	O	+0·082	NIL	+0·082
8·70 ,,	**8·80**	**P**	**NIL**	**NIL**	**NIL**
8·60 ,,	8·70	Q	−0·082	NIL	−0·082
8·50 ,,	8·60	R	−0·164	NIL	−0·164
8·40 ,,	8·50	S	−0·246	−0·200	−0·446
8·30 ,,	8·40	T	−0·328	−0·500	−0·828
8·20 ,,	8·30	U	−0·410	−0·600	−1·010
8·10 ,,	8·20	V	−0·492	−0·700	−1·192
8·00 ,,	8·10	W	−0·574	−0·800	−1·374
Less than 8·00		X	−0·656	−0·900	−1·556

Table 5.10. Average milk quality for the main breeds

Breed	Solids-not-fat	Butterfat	Total solids
Friesian	8·6	3·7	12·3
Ayrshire	8·7	3·9	12·6
Shorthorn	8·7	3·6	12·3
Guernsey	8·9	4·6	13·5
Jersey	9·1	5·1	14·2

John Nix, for example, quotes a premium for Channel Island producers of 2.25 pence over Friesians but the average yield quoted is only 3,600 litres compared to 5,000 litres for Friesians. The resultant milk sales and margin over concentrates per cow are shown in Table 5.11.

Table 5.11. Milk sales and MoCs: Friesian and Channel Island breeds

Performance level:	Friesian Average	Channel Island Average	Equal to Friesian
Milk yield (*litres*)	5,000	3,600	4,500
Milk price (*p*)	12.75	15.00	16.25
Milk sales (*£*)	683	540	731
Concentrates per litre (*kg*)	0·34	0·40	0·40
Concentrates per cow (*tonne*)	1·70	1·44	1·80
Price per tonne	140	142	142
Concentrate cost	238	204	256
Margin over concentrates	475	336	475

The above data does not show the Channel Island breeds to advantage compared to Friesians. Channel Island breed enthusiasts would probably dispute that the above was a comparison of 'like with like' and would claim a higher average milk price and higher average yield for the breed. They would also point out that the Channel Island cow is much smaller than the Friesian and that consequently more can be kept per hectare. Generally speaking, however, one has to accept that the margins per cow produced by Channel Island breeds are lower than those from Friesians. To achieve a similar margin to the average Friesian the Channel Island cow has to produce a yield of 4,500 litres and command a milk premium price of 3.50 pence per litre. Whether this can be done is debatable.

Table 5.12. Average lactation yields and butterfat percentages, England and Wales, 1978–9

	Lactation yield (*kg*)	Butterfat (*per cent*)
British-Canadian Holstein Friesian	6,218	3·76
Friesian	5,441	3·79
Ayrshire	4,863	3·94
Dairy shorthorn	4,730	3·62
Guernsey	3,941	4·65
Jersey	3,776	5·14

The average lactation yields and butterfat percentages of recorded herds in England and Wales in 1978–9 were as shown in Table 5.12. Emphasis up to now has been placed on the differences between breeds in respect of compositional milk quality. It is equally important to remember and to be aware of the differences found between families and between herds within a breed. The best Friesian herds, for example, have butterfat percentages in excess of 4·00 per cent compared to the average of 3·79 per cent.

Account also has to be taken of variations that take place through lactation and it should be remembered that good feeding and management will tend to improve compositional quality as well as yield.

SEASONALITY OF MILK PRICES

The price received for milk is also influenced by the month in which it is produced (see Table 5.13).

Table 5.13. Monthly average net prices paid to wholesale producers

	1978–9	1979–80
April	10·046	10·943
May	9·332	10·076
June	9·352	10·066
July	9·684	10·442
August	10·106	10·884
September	10·463	11·233
October	11·046	11·804
November	11·151	12·120
December	11·187	12·159
January	11·176	12·300
February	11·143	12·812
March	11·007	12·690
YEAR AVERAGE	10·430	11·407

The lowest prices occur in May and June when milk production is at a peak and cheapest to produce. The differences between the average May/June and average December/January prices in 1978–9 and 1979–80 were 1.839 and 2.158 pence per litre respectively. These differences are less than the additional cost of concentrates that need to be fed in December/January compared to May/June. Consequently one can expect higher MoCs per litre in the summer than in the winter months. *Note.* Concentrates were approximately £130 per tonne or thirteen pence per kg in 1979–80. Consequently the difference in milk price between summer and winter of 2.158 pence is only equivalent to 0·16 kg per litre.

This difference in price and in levels of concentrate feeding is one of the factors that has to be taken into account when deciding on the best months in which to calve cows. This, however, is only one of the factors that has to be considered and is usually of minor significance compared, for example, with the effect date of calving has on yield.

A switch from 50 per cent summer 50 per cent winter milk production to 40 per cent winter 60 per cent summer only changes the price received for 20 per cent of the milk produced and this change is only in the order of 1.8 pence per litre. For a cow giving 5,000 litres the change in milk sales equals 1,000 litres at 1.8p, or £18 per cow.

At 12.75 pence per litre this is only equivalent to a change in yield of 140 litres per cow.

It is also necessary to put the effects of the differential prices for compositional milk quality in perspective. An improvement in butterfat of 0·2 per cent increases the price by 0.270 pence per litre and an improvement in snf of 0·2 per cent increases the price 0.164 pence per litre. Together these give a price benefit of 0.434 pence which is worth £21.70 for a 5,000 litre cow. The same increase in milk sales can be achieved by improving the milk yield by 170 litres.

CALF OUTPUT

The calf output for a given herd is determined by the number of calves born multiplied by the price received per head after allowing for mortality.

The price received per calf is largely determined by factors outside the dairy farmer's control. It is influenced in particular by the profitability or otherwise of beef production at the time of calf disposal and whether or not there is good export trade for calves to Europe.

Calf values are subject to wide fluctuation in price from year to year, and within years, and exert a considerable external influence on the profitability of milk production. There is a tendency for prices to be low in the autumn when supply tends to be at its height and beef prices low, and high in the spring when supply tends to be low and beef prices are high, but this trend is not always consistent. It can be completely masked or exaggerated by a temporary closure of the European market or by the sudden upsurge in the demand for calves resulting from an improvement in beef cattle prices. This happened during the 1980–1 winter period. Calf prices fell to a very low level in the autumn of 1980, well below John Nix's budget figure of £75, and then increased substantially in the spring of 1981 as a result of improved beef prices.

To have calves available for sale when the market is in short supply is not easy to arrange but for those who take the trouble the rewards can be significant. Those dairy farmers, for example, who were able to retain and rear calves during the depressed autumn of 1980 to sell in the spring of 1981 have reaped a substantial reward for their efforts. It should be noted that this reward is part of the 'profit' made from calf rearing and the value of calves at birth would reflect the national picture. This together with the fact that nearly half the calves born, *i.e.*, heifer calves retained for rearing as replacements, are arbitrarily valued leads to very little difference in calf output results between farms in financial surveys.

Most of the differences in calf value that are not due to seasonal or supply and demand factors are due to differences between breeds. The low price received for Ayrshire steer calves has been a major reason for the gradual replacement of many Ayrshire cows by Friesians. The Channel Island breeds also suffer from the severe disadvantage of producing calves with a low value. Criticism is now levied at the poor calves produced by the Holstein cow and it will be interesting to see how this breed overcomes this problem in the future. Unlike the Ayrshire, however, the Holstein produces a higher yield than the Friesian and this will probably more than compensate for the disadvantage of the relatively low price of its calf.

The significance of the value and role in British Agriculture of beef from the dairy herd is not to be underestimated. The best way in which most farmers can take advantage of this is by crossing a proportion of their cows to a beef bull. Traditionally this would have been to a Hereford or possibly Aberdeen Angus but calves from European breeds such as the Charolais command higher premiums. These premiums have to be assessed in the light of the possible adverse effect of a difficult birth of a large calf on the subsequent lactation yield.

A further advantage gained by using a beef bull is the opportunity this provides to run a bull with the dairy herd towards the end of the calving season. This helps to improve the calving index, increases the number of calves born per annum, and improves the yield per cow, as well as providing calves of a higher sale value.

HERD DEPRECIATION (OR HERD MAINTENANCE COSTS)

This is shown in John Nix's budget figures at the same figure of £38 per cow irrespective of yield.

Again, like calf values, this is due in part to the fact that most dairy farmers rear their own dairy replacements and there is a tendency to transfer these into the herd at the same value per head on all farms.

A further word of explanation is necessary about the word herd depreciation. Later we shall be discussing the term 'Appreciation in Dairy Cow Unit Values' and it is more appropriate to think of herd depreciation as the 'Herd Maintenance Cost'. The cost of maintaining a typical hundred-cow herd is illustrated in Table 5.14.

Table 5.14. The cost of maintaining a typical 100-cow herd

	No.	Per head (£)	Total (£)
Cows culled	18	340	6,120
Casualties	2	40	80
	20	310	6,200
Less Purchase price or sale value of replacements	20	500	10,000
Herd maintenance cost	20	190	3,800

Clearly it can be seen that the herd maintenance cost depends on, (1) the proportion of the herd that needs to be culled, and (2) on the difference between the purchase or transfer-in price of replacements and the average price received for cull cows. As stated already, most farmers rear their own replacements so the remaining two important factors are the cull price and the proportion culled.

Note: A general point needs to be made at this stage, namely that most farmers in their costings place conservative values on the heifers they transfer into their herd. This leads to an underestimation of the herd maintenance cost and an underestimation of the gross margin produced by the dairy replacement enterprise.

The effect of this conservative value on the average dairy cow gross margin tends to be modest in percentage terms but it often has a very significant effect on the dairy replacement gross margin.

Returning to cull cow prices: these, like calf prices, tend to be high in the spring and early summer but low in the autumn and early winter when most cows are culled. Cull cow prices are also very dependent, like calf prices, on factors outside the dairy farmer's control. Again, like calf prices, they are very much influenced by the breed of cow, the large cow such as the Friesian making much more money than the smaller Ayrshire or Channel Island cow. Earlier, emphasis was placed on the value of the calf as a major reason for the present popularity of the Friesian breed. With this, one should couple the high price received for the cull cow.

What can an individual farmer do about his cull cow prices? The simple, quick answer is very little; but he can try to avoid being placed in a position of having to sell cows in poor condition and try to sell at periods of high prices. One of the advantages of the self-feed silage system, for example, is that cull cows tend to be in good condition due to their ability to put on weight on an ad-lib silage diet. If cast cows are in poor condition, particularly in mid-winter, it often pays to keep them separately on silage and/or to keep them to sell them off grass in May/June when prices are high. The margins per month to be gained from this exercise, *i.e.*, increase in value of the cow per month, can sometimes be as high as the margin over concentrates. Whether this should be done, however, does depend on the condition of the cows and the availability of feed supplies. It should also be noted that this is also one of those cases where doing the right thing for the profit of the farm as a whole can lead to a reduction in the margin per cow. The margin over concentrates per cow will be lower than it otherwise would have been if the dry cows are retained in the herd for a little longer.

This can also be true when we come to consider the other factor determining herd maintenance cost, *i.e.*, the proportion of the herd culled per year. It is generally accepted that a good calving index is necessary to achieve good margins but this can be taken to extreme by, for example, making a decision to cull any cow that is not in calf within 120 days of calving irrespective of its age or previous milk yield. The cost of replacing this cow (*i.e.*, the difference between its value as a cull and the price of a replacement) is likely to be at least £150 and possibly £200, and this represents four to five months' margin over concentrates.

The national replacement rate is in the region of 20–25 per cent and infertility or difficulty in getting in-calf is a major reason for culling. In many instances this is probably unjustifiably high for the reason given above.

One final point needs to be made about culling rates. This determines the number of heifer replacements that need to be reared per annum and in turn this can influence the number of cows that can be kept and/or the area of another enterprise such as cereals that can be grown. This indirect effect on the profit is often more significant than the direct effect on the herd maintenance cost.

Finally, as a rule of thumb it is useful to note that calf output should exceed the herd maintenance cost on a well-run dairy farm by almost as much as the sundry variable costs. This means that in effect the gross margin is simply the margin over concentrates *less* forage costs.

APPRECIATION IN DAIRY COW UNIT VALUES (AND THE HERD BASIS OF TAXATION)

This was referred to in Chapter 3 when considering standard data from costings by Manchester University.

Dairy cow unit value appreciation in this instance is the amount by which a dairy cow has increased in value between the beginning and end of the financial year. In a sense it does not really represent a profit becaue the true value of the cow has simply been adjusted in £p terms to take into account the depreciation that has taken place in the value of money.

This argument has been accepted by the Inland Revenue and the new entrant to dairy farming can choose to be assessed for tax on the 'herd basis', rather than on a trading stock basis. In the latter case the increase in the value of the cow is treated as part of profit and the farmer would be liable to tax on the valuation increase, or as he would put it, on his 'paper profit'. If he adopts the herd basis this increase in valuation is not taxed. More important, if and when the herd is sold there is no tax payable on the difference between the actual sale value and the original cost of the herd. This is of particular significance in times of inflation such as those that have occurred during the 1970s. During this ten-year period the cost of a dairy cow increased approximately four-fold, *i.e.*, from £100–125 to £400–500 per head.

Investment in dairy cows, like land, is in a sense a hedge against inflation. The increase in cow values, for example, between 1970 and 1980 was roughly in line with inflation. It should be noted, however, that the increase in cow values is not always closely in step with increases in the general level of inflation; it tends to lag behind.

The fact that investment in dairy cows may provide a hedge against inflation needs to be borne in mind when comparing dairy farm profits with arable farming, particularly if dairy farming profits are at a low ebb. When dairy farm profits improve the dairy farmer benefits both from this and from an increase in cow values.

SUNDRY VARIABLE COSTS

These include bedding, veterinary services and medicines, AI and recording fees, and consumable stores.

The significance of bedding costs depends on the location of the farm. In the wetter all-grassland areas it is a significant item but in the midlands and eastern counties it is a cost that can be largely ignored because bedding supplies are plentiful and cheap.

John Nix assumes that very little change takes place in these sundry costs as yields rise and this is borne out by the survey data.

Care has to be taken to control these costs like any other costs on the farm but this is not a crucial factor.

Perhaps the most important single item is that of veterinary services and medicine. This cost is minimal on some farms where there is little veterinary work other than the routine visit for everyday fertility and disease problems.

As herds are moved towards higher levels of production, so there is a greater need for veterinary involvement in maintaining herd health. Pregnancy diagnosis is becoming a regular feature on many farms. This development, together with the increasing cost of drugs, can result in higher veterinary charges in high yield herds.

FORAGE COSTS

These represent the variable costs incurred to provide grazing and winter feed for the dairy cow and include seeds, fertilisers, sprays and crop sundry items, such as silage additives.

Contract charges for silage making and grass keep taken are sometimes included as part of forage costs. They are variable costs according to the definition of that term but for comparative purposes it is much better if these items are included as part of fixed costs. Purchased forage feeds such as hay are also included sometimes as part of forage but it is much better if they are included under a heading separate from both concentrates and forage as 'Other Purchased or Bulky Feed'.

The interpretation of forage costs data is one of the most difficult exercises in farm business management, as they are influenced by so many factors. When looking at the forage costs one has to try to take into account so many elements at the same time, such as the inherent quality of the land, the stocking rate, system of grassland conservation, the level of concentrate feeding and milk yield. In addition, one has to make this interpretation knowing that the forage costs have probably been allocated to the individual enterprises on a livestock unit basis and this allocation may be inaccurate. The first step in any analysis of forage cost data should therefore be to examine these in relation to the grazing livestock enterprises as a whole, **and then and only then** in relation to the individual enterprises.

A final point to remember is that many farmers' knowledge of the precise number of hectares they farm is hazy. The area of cash crops grown is usually known fairly accurately and the area of forage is then deduced by difference, *i.e.*, by subtracting the area of cash crops from the total farm area, including buildings and roads. Beware, therefore, of any figures pertaining to stocking rates; these may or may not be accurate! The same point applies to forage costs, accurate records are often kept of fertilisers used on cash crops and the 'difference' *i.e.*, total less these is then charged to grassland!

NATIONAL ECONOMIC AND POLITICAL FACTORS

The economic and political factors of this country, and in more recent times of the other EEC countries, influence and determine the price the dairy farmer receives for his milk, the price he pays for purchased feeds, and the cost of other inputs such as labour and machinery. The individual farmer has very little influence on these factors and there is relatively little he can do about them.

It has to be stressed, however, that the success or otherwise of government policy and the influence or otherwise dairy farmers as a whole have on this policy through organisations such as the National Farmers Union has a very considerable effect on the profitability of milk production.

Historically milk production has been one of the most profitable enterprises in British farming. Generally it is regarded as more profitable than beef and sheep production and it accounts for a very significant proportion of the total national farm output.

Prior to entry into EEC in 1973 the price of milk, and hence the profitability of dairy farming, was very dependent on the proportion of milk being used in the liquid (*i.e.*, retail milk market rather than the less profitable manufactured market, particularly butter). Manufactured milk could be, and was, imported at low prices. Consequently any substantial expansion in milk production above the liquid market requirements led to a significant fall in producers' returns. Government policy at that time was largely based on the liquid milk market and the concept of 'Standard Quantities' became established. If production exceeded the standard quantity the price received by producers fell due to the low price received for manufactured milk.

Since our entry into EEC the import of milk products at low world market prices has not been possible due to the EEC system of levies on imports. The price received for milk used for manufacture has increased and this has allowed the home production of milk to be increased without any substantial fall in prices.

These trends are illustrated by information taken from *Dairy Facts and Figures 1980*, a publication produced annually by the Milk Marketing Board (see Tables 5.15 and 5.16).

Over the 25-year period 1954–5 to 1979–80 there has been very little increase in the amount of milk sold liquid. Liquid sales peaked in 1974–5 and have fallen slightly during the past five years.

Total production, however, has continued to increase and manufactured milk sales now account for 50 per cent of production compared to only 19 per cent twenty-five years ago.

Table 5.15. Utilisation of milk produced off farms in England and Wales

Year	Liquid (Million litres)	(per cent)	Manufacturing (Million litres)	(per cent)	Total (Million litres)
1954–5	6,103	81	1,413	19	7,516
1959–60	6,253	77	1,923	23	8,175
1964–5	6,647	73	2,398	27	9,044
1969–70	6,664	66	3,358	34	10,022
1974–5	6,878	62	4,273	38	11,115
1979–80	6,432	50	6,342	50	12,774

Due to the change in market support arrangement the increase in the price received for manufactured milk between 1969–70 and 1979–80 was 553 per cent, whereas the increase in the average net price received by producers for all milk supplies was only 335 per cent.

Table 5.16. Changes in net price received for milk

Year	Average net price received by wholesale producers (p/litre)	Average price received for manufactured milk (p/litre)
1964–5	3.403	2.038
1969–70	3.548	1.992
1974–5	6.245	5.017
1979–80	11.407	11.026
	per cent	per cent
1979–80 as per cent of 1969–70	335	553

A similar expansion in output has taken place throughout the EEC and by the end of the 1970s the EEC as a whole had a surplus of milk products. These surpluses are now an embarrassment and their removal is one of the main problems facing policy-makers in the 1980s.

Although the EEC as a whole is in surplus this is not the case so far as the British market is concerned. British farms only produce about 70 per cent of the national requirements, 30 per cent still being imported. Whether or not this shortfall in home supplies is to be met by an expansion of home production or by imports is the centre of political debate at the present time. The answers have important repercussions for British dairy farming during the 1980s and will largely determine its profitability relative to other enterprises.

The possible continued change in the market outlet, *i.e.*, towards a greater proportion going into manufacture, also has implications from the point of view of milk compositional quality. As the proportion going into manufacture increases so does the significance of the butterfat and solids-not-fat content of the milk. This led to the introduction by the Milk Marketing Board of a new quality payments system described earlier in this chapter.

Considerable stress is placed at the present time (1981) on the need for farmers to pay more attention to marketing. For many years dairy farmers have been fortunate in being able to concentrate on the job of producing milk and leaving the job of marketing to the Milk Marketing Board.

Hopefully this state of affairs will continue in the future but the milk producer must heed the market and price signals coming from the Board. These are expected to place more emphasis on the need for high butterfat and possibly less emphasis on winter milk production. The producer needs to take note of these signals and alter his production methods accordingly.

Chapter 6

DAIRY REPLACEMENTS

ROLE IN THE WHOLE FARM ECONOMY

Attention was drawn in Chapter 2 to the three main relationships between enterprises: competitive, complementary and supplementary. Emphasis was placed on the need to select a combination of enterprises and a system of farming that leads to a well-balanced system. Dairy replacements are now considered with these principles in mind.

The profits made rearing dairy replacements are nearly always considerably less than those from dairy cows. This is true whether the return is measured per hectare, per £100 labour or per £100 capital.

It follows that as a general rule the number of dairy replacements relative to the number of cows should be kept as low as possible. Too many dairy heifers relative to dairy cows is a common weakness on many dairy farms and is usually due to either a high rate of culling from the dairy herd and/or too long a period between birth and calving. On most well managed dairy farms, heifer numbers should not exceed 50–60 per cent of the dairy herd numbers.

Despite their low relative profitability there are very few dairy farms on which the rearing of some dairy heifers is not justified and over two-thirds of dairy farms supply all their own replacements. Farms on which heifer rearing is *not* justified are likely to have most of the following characteristics:

- No land that is not suitable for either grazing by the dairy herd, conservation or for growing arable crops.
- A severe shortage of working capital.
- Adequate buildings and equipment to house all the cows required to utilise the existing grassland and forage area.
- A farm staff that can be kept fully employed and motivated without the interest of rearing replacements.
- A manager or farmer capable of buying heifers as good as those he could rear on the farm.

Dairy heifer rearing is justified on a farm where some or all of the above conditions are not met. Most farms have an area of permanent pasture that is too far away from the buildings to be grazed by dairy cows, and this in itself often justifies heifer rearing. On many farms this area is much larger than the requirements of the 50–60 per cent dairy heifers mentioned above. This leads to a high proportion of dairy heifers to cows and to a lower level of actual profit per hectare. In this case thought needs to be given to having an autumn-calving herd so there are dry cows to utilise this area during most of the summer.

The provision of winter feed for young stock usually competes with the needs of the dairy herd but there is a place for young stock in the sense that their requirements for winter food are not as critical. Alternative food such as straw can be used in times of shortage, leaving all the silage for the dairy herd. Young stock can also act as scavengers utilising leftovers from the dairy herd, but this strategy conflicts with the need to rear the heifers to calve at an early age.

PROFITABILITY IN THE WHOLE FARM ECONOMY

As dairy heifer rearing is basically unprofitable the main objective should be to manage this enterprise in such a way as to make effective use of the resources not required by the dairy herd.

Two-year calving is necessary to get high gross margins per head from a young stock enterprise but the achievement of this objective may not be feasible if building resources are limited and may not be desirable if outwintering is a feasible proposition. In this instance it may be more prudent to calve at two and half years of age and adjust winter feed costs accordingly.

Although it is emphasised that the rearing of dairy heifers is relatively unprofitable this degree of unprofitability is often over-estimated. This is due to two basic errors in most costing systems. Firstly, the calf born to the heifer is credited to the dairy herd, *not* to the heifer that produced it, and the newly calved heifer is usually transferred to the dairy herd at too conservative a value relative to the cost of purchasing newly calved replacements. Secondly, forage costs are usually charged to the young stock on a livestock unit basis. This system results in the young stock being overcharged as they graze the less productive pastures and normally require and obtain less conserved forage than their livestock unit equiv-alents would suggest. When assessing the contribution of your young-stock enterprise, therefore, please try to ensure that the allocation of forage acres and costs is done on the basis of grazing records and known levels of conserved feed utilisation.

The effect of these two errors in recording on the gross margin obtained from a young stock enterprise is illustrated in Table 6.1.

The young stock enterprise in Example A shows a gross margin of £2,888 or £48 per head. In Example B this is increased to £76.8 per head or by 60 per cent. This is achieved simply by adding £45 to the value of the heifers so as to include the value of its calf and by assuming a more accurate allocation of the forage costs.

Table 6.1. Two examples of dairy replacement gross margins

Ave. no. replacements				Example A 60		Example B 60	
					£		£
Heifers in hand at end of year				60 @ £250	15,000	ditto	15,000
Casualties				2		ditto	
Heifers transferred to dairy herd				22 @ £500	11,000	22 @ £545	11,990
Heifers culled				2	500	ditto	500
SUB TOTAL				86	26,500	ditto	27,490
less calves transferred from dairy herd				26 @ £50	1,300	ditto	1,300
plus heifers on hand at start of year				60 @ £250	15,000	ditto	15,000
				86	16,300	ditto	16,300
GROSS OUTPUT					10,200		11,190
Variable Costs				£ per head	£	£ per head	£
Feeding stuffs				70	4,200	ditto	4,200
Veterinary fees and medicines				5	300	ditto	300
Sundries				5	300	ditto	300
Forage costs				42	2,512	30	1,800
				122	7,312	110	6,400
GROSS MARGIN				48	2,888	76.5	4,590

Forage Costs Allocation					
					£
				Young stock	
Livestock units	No.	Factor	No. units	grazing based on	
Heifers over 2 years	12	0.4	4.8	grazing records:	1,000
Heifers 1–2 years	24	0.6	14.4	Conserved forage	
Heifers under 1 year	24	0.8	19.2	based on forage	
				utilisation recds:	800
Total Heifers	60	0.64	38.4		
Dairy cows	90	1.0	90.0	Total to heifers	1,800
				Total to cows	
Total Livestock Units			128.4	(by difference):	6,600
TOTAL FORAGE COSTS			£8,400		8,400

Forage costs per livestock unit (8,400 ÷ 128.4) = £65.42
Forage costs allocated to dairy heifers 38.4 × 65.42 = £2,512
Forage costs allocated to dairy cows 70 × 65.42 = £5,888

The next conventional step in the analysis of the dairy replacements is to calculate the area of land used by the replacement, again using the livestock unit basis and then to determine the gross margin per hectare and compare this to that achieved by the cows. However, on most farms this is rather a pointless exercise as there is usually no similarity between the areas of land occupied by the cows on the one hand and young stock on the other. The former occupy the best grazing acres and utilise the best conserved forage. The latter occupy the worst grazing land and utilise whatever conserved forage is available after the needs of the dairy herd have been met.

Having stressed the perils of using conventional data to assess dairy heifer performance we need to turn back to means of measuring and improving efficiency.

AGE AT CALVING

Age at calving, as already mentioned, is a most vital factor. A low age of calving needs to be achieved consistent with the dairy heifer enterprise fitting into the farm system after the needs of the dairy herd have been met.

We need to be aware, however, that lowering the age of calving does not necessarily lead to any significant direct benefits resulting from improvements in the profitability of the young stock themselves. Most of the benefits tend to be indirect. This is illustrated by calculating the gross margins before forage that can be expected from heifers reared to calve at (a) 2 years to 2 years 3 months, and (b) 2 years 9 months to 3 years (see Table 6.2).

Lowering the age of calving increases the gross output per annum by £43 but this is cancelled out by a similar increase in feed costs. Consequently the gross margin before forage costs tends to be similar for both systems when considered on a 'time basis', *i.e.*, the gross margin per head per annum tends to be the same irrespective of the age of calving for a given level of management efficiency. In practice, however, better managers achieve earlier calving at no greater feed costs than their less efficient neighbours due to their better stockmanship and greater skill in utilising grassland.

Lowering the age of calving becomes much more significant, however, when we look at the land required (see Table 6.3). Calving at 33–36 months increases the area of land required by approximately 50 per cent compared to calving 24–27 months.

The most significant factor is the 50 per cent saving that can

Table 6.2. Gross margin before forage costs at two ages of calving

Age at calving	2 years to 2 years 3 months (e.g., born Oct–Nov, calve Oct–Feb)		2 years 9 months to 3 years (e.g., born Oct-Nov calve Jul–Nov)	
		£		£
Value of heifer		545		545
Less value as a calf (50 + 6 per cent mortality)		53		53
Gross output		492		492
Variable Costs				
Feed (a) *0–6 months*				
Milk substitute 13 kg @ 60p		8		8
Concentrates 350 kg @ 14p		49	300 kg @ 14p	42
(b) *6–12 months* (May–October)	250 kg @ 12p	30	100 kg @ 12p	12
(c) *12–18 months* (Nov–April)	600 kg @ 12p	72	250 kg @ 12p	30
(d) *18–26 months* (May–Dec)	100 kg @ 12p	12	Nil	—
(e) *26–30 months* (Jan–April)			150 kg @ 12p	18
(f) *30–34 months* (May–Nov)		—		—
Total feed		171		110
Vet fees and sundries		26		26
Total variable costs		197		136
GROSS MARGIN BEFORE FORAGE COSTS		295		356
Age at calving (months)		26		32
		£		£
Gross ouput per month		18.9		15.4
Gross output per year		227		184
Gross margin before forage costs per month		11.35		11.25
Gross margin before forage costs per year		136		133.5

be made in winter feed, or hay equivalent requirement. This land is released for another enterprise and the real or opportunity cost of calving at a later date is the profit that could be generated from this land by the alternative enterprise. The area of conservation land that can be saved by lowering the age of calving is in the region of 0·2 hectare per heifer reared. Devoted to cereals this land would give a gross margin of £400–£500 per hectare or £80–£100 per heifer reared.

Table 6.3. Age of calving and land requirement

Age at calving Forage requirements		2 years–2 years 3 months Hectare High nitrogen	Low nitrogen		2 years 9 months–3 years Hectare High nitrogen	Low nitrogen
Grazing: first summer		0·10	0·15		0·10	0·15
second summer		0·20	0·30		0·15	0·22
third summer					0·20	0·30
		0·30	0·45		0·45	0·67
Hay equivalent:	(tonne)			(tonne)		
first winter	0·6	0·13	0·20	0·5	0·11	0·17
second winter	0·9	0·20	0·30	0·75	0·17	0·25
third winter				1·00	0·22	0·33
	1·5	0·33	0·50	2·25	0·50	0·75
TOTAL		0·63	0·95		0·95	1·42

The opportunity cost of the 50 per cent saving in the area required for grazing could be much less significant on many farms where there is a substantial area of permanent grass not suitable for grazing by the dairy herd. The saving may simply be a reduction in fertiliser costs resulting from the lower stocking density.

A second consequential benefit of earlier calving is the reduction in building requirements and the labour required to tend the stock in winter. It is difficult to quantify this saving because once again it will depend on the alternative use of the buildings and labour.

Finally, there is a substantial cash flow benefit from earlier calving. This is particularly important where capital is limited or expansion is planned or needed in dairy cow numbers.

By spending an additional £60 or thereabouts per head on feed it is possible to bring forward the date of calving of a group of heifers by 9–12 months. The total value of the calf plus the margin over concentrates produced in this period is in the region of £440. The net effect on receipts and payments known as cash flow is shown in Table 6.4.

The effect on the gross margin of an established business of calving at 24–26 months of age instead of at 34–36 months is illustrated in Table 6.5.

Calving at an earlier age releases six hectares of land and this will allow twelve additional cows to be carried providing there are

Table 6.4. **Effect on receipts and payments of earlier calving**

Period/age (months)	6–12	12–18	18–24	24–30	30–36	36–42
Additional receipts	£	£	£	£	£	£
Calf	—	—	—	40	—	—
Milk	—	—	—	100	300	200
				140	300	200
Less feed payments	18	42	12	50	120	80
	(18)	(42)	(12)	90	180	120
ACCUMULATED SURPLUS (DEFICIT)		(60)	(72)	18	198	318

adequate buildings and presuming that the land vacated is suitable for dairy cows.

The increase in the gross margin these cows would generate may be the difference between a worthwhile profit, breaking even or making a loss. Note that the economic contribution made by improving the young stock enterprise is hidden in the sense that the young stock enterprise shows no improvement in its own gross margin; in fact the gross margin per head is down due to the need for more supplementary feed.

Table 6.5. **Effect on gross margin of earlier calving**

Calving age	24–26 months			34–36 months		
Gross margin	No.	£	hectares	No.	£	hectares
	112 @ £350 =	39,200	56	100 @ 350 =	35,000	50
	28 @ £250 =	7,000	19	25 @ 280 =	7,000	25
TOTAL		46,200	75		42,000	75

ORGANISING MORE PROFITABLE HEIFER REARING

Having looked at length at the profitability of replacements in terms of their place on the farm, we now turn to examine other ways in which their contribution can be increased.

1. By rearing heifers of intrinsically better value. The cost of rearing a heifer of low genetic potential is no less than one with

a high potential. It is important, therefore, to take care over the breeding policy. This must be kept in perspective, however, as management and feeding have just as much effect if not more on yields and margins from dairy cows. The cost of nominated services per heifer *eventually* reared is higher than necessary on many farms.

2. By rearing heifers born at the time that favours early calving, *i.e.*, heifers born in the autumn rather than in the spring. The late winter and spring-born calf is not large enough to utilise grazed grass effectively in its first year and is difficult to manage to calve at two years without incurring very substantial feed costs.

The best way to achieve this objective in many herds is by using a beef bull on the dairy herd from the end of February or March until the end of September. This may necessitate some sacrifices in genetic potential but in most instances this is outweighed by the next factor.

3. By rearing heifers born over as small a time scale as possible so as to provide even bunches of cattle for rearing, preferably one bunch. This facilitates the adoption of a controlled rearing pro-gramme designed to bring the heifers to the correct service weight which in the case of Friesians is 330 kg by the time they are 15–18 months of age (see Fig. 4). One of the biggest problems in heifer rearing is the wide age range often found in groups of cattle running together. The objective is to try to adopt a similar strategy to that favoured by efficient '18 month beef' producers.

4. If possible, plan the main calving data so that it fits in with the needs of the rest of the farm as well as the dairy herd. The ideal time to start from the dairy herdsman's point of view may be the beginning of September but if cereals are important to the farm economy it may be better to delay the start until the end of September.

Further delay, however, is not desirable unless the calf-rearing facilities are excellent.

5. Organise a work routine and calf-rearing policy that fits in with the facilities available. Have spare capacity to deal with surplus steer calves so these can be retained in times of surplus.

Individual bucket feeding of calves is still regarded by many as the best method for keeping both mortality and feed costs down, but labour availability may necessitate a more streamlined and mechanised feeding system, even if this does mean an increase in feed costs per head.

6. Aim at a liveweight of 180 kg before turn-out in April. Calves of this weight or more make effective use of grazed grass. Try to

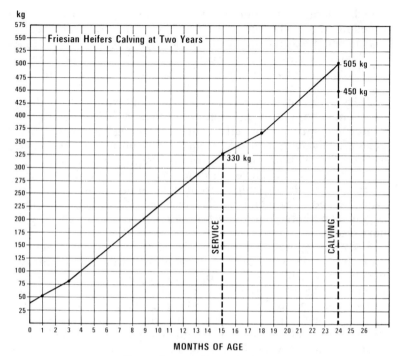

Fig. 4. Target Growth Curve

turn the calves out before there is abundant grass available to avoid digestive upsets.

7. Plan the first summer grazing strategy for the young calves. Ideally this will include a move to clean aftermath in late July, possibly after second-cut silage, with the objective of minimising worm infestations and maintaining liveweight gain.

8. Introduce supplementary concentrate feeding in late summer if necessary to reach target weight at yarding.

On light land some economy in winter feed and concentrate costs may be feasible by growing kale and/or stubble turnips.

9. Bring cattle indoors in good time and if necessary give supplementary feed to reach target service weight. If cattle have to be outwintered don't adopt a 'cheap feed policy at all costs' attitude—feed to reach target weights.

10. Plan the grazing policy for the following season with the same precision as you would that for the cows.

11. To sum up, the profitability of this enterprise is increased by giving as much thought as possible to its planning and day-to-day management. Dairy heifer rearing is a neglected enterprise and contributes less than it should to profits on nearly all farms.

REARING DAIRY HEIFERS FOR SALE

If we decide to rear heifers surplus to replacement requirements for sale we need to consider whether this is more profitable than an alternative enterprise. In considering this decision it is presumed that the needs of the dairy herd have already been met and that expansion of dairy cows is not justified.

The possible alternatives to dairy heifers will depend on the nature of the land available. If this is mainly permanent pasture then the comparison in profitability is most likely to be between dairy heifer rearing and beef cattle. In practice there is relatively little difference in the profitability of these two alternatives and a decision can usually be made based on personal preference. If, however, the land can be used to grow cereals, it is most likely that a partial budget would show significantly in favour of growing cereals rather than keeping more young stock. A decision to rear heifers in this case should be taken with the knowledge of the real cost to the farm economy of personal preference.

The case for rearing dairy heifers for sale rests largely on the price that can be obtained for the heifers. If a premium can be obtained over and above that achieved by the average farmer it is more likely to be a profitable alternative to cereals and other livestock enterprises. This is most likely to be achieved by a pedigree herd and the place of heifer rearing in the pedigree herd is now considered in rather more detail.

PEDIGREE BREEDING AND REARING

At the outset it is probably as well to define a pedigree herd. There are many commercial herds containing pedigree cattle but the main objective of the herds is to produce milk. A pedigree dairy herd is really a herd whose purpose is to produce heifers for sale and where milk is a by-product.

The economy of these farms and the relationship between the profitability of milk production on the one hand and the production of heifers on the other is fundamentally different from that on the commercial farm.

Firstly, the cows being used to produce milk have a much higher

Table 6.6. Grazing livestock gross margin budgets for a pedigree herd (1981–2 price levels)

Average No: Cows: 100 Other cattle: 150	Dairy herd (No.)	£	Per cow £	Dairy heifers (No.)	£	Total £
Closing Valn.	100	80,000	800	150	75,000	155,000
Stock Sales						
Cull cows/heifers	14	4,900		2	800	5,700
Casualty	1	—		2	—	
Surplus heifers				50	40,000	40,000
Steer calves	50	4,000		25	1,100	5,100
Bulls				4	5,000	5,000
Milk sales		86,400	864	—	—	86,400
Transfers out	50	4,000		15	12,000	16,000
Sub-total (A)		179,300		248*	133,900	313,200
Opening Valn.	100	80,000	800	150	75,000	155,000
Stock Purchases						
Transfers in	15	12,000		50	4,000	16,000
Sub-total (B)		92,000		200	79,000	171,000
				per head		
GROSS OUTPUT (A−B)		87,300	873	366†	54,900	141,200
Variable Costs						
Bought concentrates		34,400	344	60	9,000	43,400
Other bought feed		—	—			
Home-grown grain		—	—			
Vet & medicines		2,500	25	6	900	3,400
Other items		4,400	44	5	750	5,150
Total ex-forage		41,300	413	71	10,650	51,150
Gross margin ex-forage		46,000	460	295	44,250	90,250
Forage costs		7,000	70	35	5,250	12,250
GROSS MARGIN		39,000	390	260	39,000	78,000

Milk yield per cow (litres) 6,400
Milk price per litre 13·50
MoC per cow 520
* Including 48 calves born to the 50 heifers not transferred into the dairy herd
† Gross output per head (total ÷ ave no.)

opportunity cost, *i.e.*, sale value, than that found on commercial farms. The average sale value will depend on the herd and its reputation. If it is a true pedigree herd as defined above, one would expect the sale value of heifers to be at least 150 per cent of normal prices, *i.e.*, £800 or more compared to £500–550. The sale value of elite cows in the herd is probably at least four times that of a typical commercial cow.

The gross margin that could be expected from a pedigree herd is shown in the budget in Table 6.6. Attention is drawn to a comparison of the figures in this budget with those prepared for the 80 hectare case study commercial farm described in Chapter 10.

The case study farm is assumed to keep 140 cows and to rear 42 heifers per annum, *i.e.*, just enough to ensure adequate replacements for the dairy herd and leave a small surplus for sale. The pedigree breeder on the other hand would want to rear all the heifers born to the dairy cows and to the heifers. At a similar stocking density he is able to keep 100 cows and rear about 70 heifers and 4 bulls per annum.

The budget for the pedigree farm shows a similar total gross margin to that for the commercial farm *even though* the margin over concentrates is only budgeted at £520 per cow compared to £570 on the commercial farm. This similar total gross margin is dependent, however, on there being fifty surplus heifers for sale at an average price of £800 and four bulls at an average of £1,250.

The capital 'tied up' in livestock on the pedigree farm is about twice that of the commercial farm but labour and machinery costs may be lower due to lower cow numbers. Whether or not one can make more from pedigree farms depends, therefore, on being able to command a high price for heifers. In turn, this largely depends on the quality of the basic foundation stock and on the capital available to purchase and retain these animals.

Chapter 7

OTHER ENTERPRISES ON THE DAIRY FARM

RELATIVE PROFITABILITY

An indication of the gross margins that can be expected from enterprises frequently found on a dairy farm is shown in Table 7.1. These estimations assume that the enterprise is managed at average to above-average efficiency.

Table 7.1. Relative gross margins of enterprises found on dairy farms

Enterprise	Gross margin £ per head	number per hectare	Gross margins per hectare £ (1981 costs/prices)	Marginal capital requirement £ per hectare
Potatoes			800–1,200	1,400
Dairy cows	320–400	2·5	800–1,000	1,250
Dairy cows	320–400	2·0	640–800	1,000
Sugar beet			600–800	1,000
Winter wheat			350–500	300
Spring barley and other cereals			300–400	250
Dairy replacements	80–100	4·0	320–400	1,000
Dairy replacements	80–100	3·0	240–300	750
Beef cattle		ditto dairy replacements		
Breeding ewes	25–30	10	250–300	500
Breeding ewes	25–30	8	200–240	400

The enterprises are ranked in their approximate order of profitability per hectare. This order will, of course, vary according to the suitability of the land for the enterprise concerned.

Return per hectare is not the only criterion. The labour required and capital requirement also have to be taken into account. An indication of the marginal or working captial required per hectare is given alongside the gross margin. Attention is drawn to the low working capital requirement for cereals.

100

Note: the working capital requirement for a cereal enterprise is roughly equal to one year's variable costs plus one year's fixed costs. It does not include fixed capital required for machinery and buildings.

CEREALS ON THE DAIRY FARM

Cereals are found and can be justified on the majority of dairy farms for the following reasons:

1. Properly managed they produce a gross margin per hectare that is higher than that obtained from grazing livestock other than dairy cows unless the grazing livestock is very intensively managed.

2. The working capital required to grow cereals is much less than that required for grazing livestock. Two to three hectares of cereals can be grown with no more working capital than that required for one hectare devoted to livestock.

3. They reduce the costs of bedding. *Note:* When you assess the gross margin of a cereal enterprise on a dairy farm you should include the value of the straw they produce in the gross ouput as well as the value of the grain.

4. They make it easier to reseed leys and facilitate the disposal of slurry and farmyard manure. Worn-out leys, for example, can be used to dispose of farmyard manure during the atumn before they are ploughed for wheat. Alternatively, slurry can be applied to the stubble during the winter before reseeding in the spring.

5. A higher stocking rate can be adopted with less fear of being short of fodder in a drought year due to the availability of feeding straw.

6. Savings can be made in purchased concentrate costs by making use of home-grown grain. It is normal practice to charge grain fed to livestock at the price it would have realised if sold, but the real saving to the dairy farmer is what it would have cost to *purchase* cereals. This is some £4–6 per tonne more than the sale price due to transport costs, etc.

7. Cereals can be grown on the dairy farm with relatively low fertiliser costs because advantage can be taken of the build-up in fertility via leys and manure.

An indication of the gross margins produced by a wheat enterprise at average and above average levels of efficiency is shown in Table 7.2.

These figures demonstrate the importance of high yields in obtaining good gross margins from cereals. To achieve these high

Table 7.2. **Wheat enterprise gross margins**

		Average per hectare £		Above average per hectare £
Grain output	5·0 tonne @ £100	500	7·0 tonne @ £100	700
Straw	2·0 tonne @ £20	40	2·5 tonne @ £20	50
		540		750
Variable Costs				
Seed		35		35
Fertilisers		65		65
Sprays		60		60
		160		160
GROSS MARGIN		380		590

gross margins it is necessary to drill the wheat crop early, that is, by the middle of October.

If winter wheat follows a ley this will need to be ploughed no later than the end of September and the utilisation strategy for the ley should be planned accordingly. This may conflict with needs of the livestock enterprise such as late autumn grazing. The dairy enterprise manager may instinctively make the wrong decision in this instance by automatically giving preference to the needs of the livestock even to the extent of sacrificing not only yield but also the winter wheat crop in favour of spring barley. The opportunity cost of this decision to grow a spring barley crop instead of winter wheat is likely to be in the region of £100 per hectare (£450 per hectare wheat gross margin less £350 per hectare barley gross margin.)

The discussion to date has centred round the gross margin contribution a cereal enterprise can make when introduced on to a dairy farm as an enterprise in its own right. A feed substitution case for cereals can also be made: that it is more economical to devote some land to cereals to feed to cows than to grow grass to conserve as hay or silage. This will not often be valid in the wetter far western half of the country but it is often true in the midlands and eastern counties. The case for grass in favour of cereals rests largely on its ability to produce higher quantities of dry matter per hectare when large quantities of nitrogenous fertiliser are used.

Some idea of how high this yield has to be can be gauged by calculating the yields of silage and hay that have to be produced to give the same gross output as that attainable from wheat (see Table 7.3).

Table 7.3. Output per hectare of wheat, hay and silage

	Average wheat crop	Above average wheat crop
Output per hectare	£540	£750
Equivalent output as hay	10 tonne @ £54 per tonne or	14·4 tonne @ £54
	12 tonne @ £45 per tonne	16·7 tonne @ £45
Equivalent output as silage	25 tonne @ £21·7 per tonne	34·6 tonne @ £21.7
	30 tonne @ £18 per tonne	41·7 tonne @ £18

One has then to turn to the question how much does it cost to produce these forage yields compared to growing wheat? Variable costs are likely to be very similar, savings in seed and spray costs required to grow grass being offset by additional fertiliser requirements. Fixed costs are difficult to assess and will depend on the individual farm circumstances but it is fair to say that the harvesting costs of cereals are likely to be much less than those for silage.

This kind of questioning approach to feed substitutes has recently been adopted by farmers who have changed to complete diet feeding. As a result such farmers now include a greater proportion of cereals in the diet but it needs to be stressed that this substitution can only be done within strict nutritional limits due to the cow's basic need for long fibre.

One of the interesting problems arising from this kind of thinking is the need to adopt different means of measuring feed utilisation efficiency. A farmer substituting half a tonne of cereals for 2·5 to 3·0 tonne of silage will reduce the margin over concentrates by £50 per cow, (the sale value of the cereals), *but* the profits of the whole farm will increase if the cereals are produced at a lower cost than the silage.

POTATOES

Potatoes produce a high gross margin per hectare and are an effective way of improving dairy farm profits given the right climate and soil conditions.

This is particularly true, for example, in certain parts of Devon and Cornwall, Pembrokeshire and Cheshire where the climate is ideal for early potatoes. Dairy cows and potatoes fit together very well on these farms because double cropping can be practised: for instance, early potatoes followed by kale or subble turnips. Advantage can also be taken of the benefits farmyard manure has on potato yields, although these benefits are questioned by some experts.

Potatoes are a better choice than cereals—given that both crops can be grown well in the area—on dairy farms that are relatively overstaffed, for instance the family farm needing to increase its output when the sons return home from college.

Potatoes, however, are a specialist crop which require substantial amounts of capital per hectare. Whether they should be introduced will depend on the availability of capital and whether there are buildings on the farm that can be readily adapted for potatoes.

This availabililty, or otherwise, of buildings applies equally to cereals. There is usually a substantial amount of buildings on a dairy farm but a large proportion tend to be specialised. General-purpose buildings, however, can be used for both cattle and cash crops. Cereals, for example, can be stored until November and then sold to make room to house young stock.

SUGAR BEET

Sugar beet is found on farms keeping dairy cows but on such farms it is probably more a case of dairy cows being kept on an arable farm rather than beet on the dairy farm.

To produce a good profit sugar beet needs to yield in excess of 35 tonne per hectare and these yields can only be achieved year in year out on grade I and grade II land. Sugar beet, however, is grown on grade III land and it is on such farms that it is more likely to be associated with milk production. Yields on such farms will tend to be below the national average and effective use will need to be made of sugar beet by-products to justify the crop continuing to be grown on the farm.

The by-products take two forms: beet tops and beet pulp. The latter can be purchased by any dairy farmer through a merchant but the beet grower is able to purchase at a lower price, usually in the region of £4–6 per tonne. Although sugar beet pulp is very good feed for dairy cows the advantage gained by being able to purchase at a lower price is not in itself very significant.

Beet tops have considerably more potential value but simple means of using these effectively in the rations of dairy cows are difficult to devise. Some arable farms have made reasonably quality silage and dry cows can make good use of beet tops in late October to early December. They can also be used to reduce the cost of feed for dairy replacements.

The yield of beet tops is in the region of 20–25 tonne per hectare. In certain instances these can be used to save the equivalent of 5·8 tonne of silage or two to three tonne of hay, representing a financial saving in the region of £100 per hectare. This kind of saving makes the difference between sugar beet being justified or not justified on the farm as a whole. Note that £100 is equivalent to approximately four tonne of sugar beet which in turn is equivalent to approximately 11 per cent of the average national yield per hectare.

CALF REARING

A consideration of the place of the cattle on the farm starts with calf rearing. Calf rearing is competitive for labour and capital but not for land. The rearing of surplus calves to three–six months of age is a means of increasing the overall farm profit if labour and buildings are available.

Calf prices fluctuate considerably from year to year as well as seasonally, and the ability to rear calves in a period of depressed calf prices lends stability to a dairy farm business. The case for establishing a specialised calf-rearing enterprise where surplus labour and buildings are not available is, however, difficult to make.

Having reared calves the temptation is to keep them too long. As a general rule any calves born during the winter period should be sold before the end of the following June so as to obtain the high seasonal prices. This will be good for cash flow because it coincides with the time by which fertilisers should have been paid for, and on a tenanted farm this follows closely after the need to find cash to pay the rent bill.

It is very difficult to establish what contribution calf rearing makes to overall farm profits because calf prices vary enormously from week to week. Calves are sold as 'calves' whether they are a few days old or two to three months of age.

It is up to the individual manager to try to determine the price he can get for week-old calves in his locality, add the cost of rearing and then make a judgement as to whether the sale price is likely to justify the additional time and effort. The latter will certainly be needed but is not likely to be reflected in any significant increase in fixed costs; the main costs he has to assess are the variable costs of rearing the calves. An indication of what these are likely to be is given in Table 7.4.

Table 7.4. Main variable costs of calf-rearing

	Rearing to 3 months of age £ per calf		Rearing to 6–8 months of age £ per calf
Milk substitute 13 kg @ 60p.	8	ditto	8
Other concentrates 15 kg @ 16p	24	ditto	24
Additional concentrates		200–260 kg @ 14p	28–36
Total concentrate feed	32		60–68
Opportunity cost of hay 60 kg @ 5p	3	200–300 kg @ 5p	10–15
	35		70–83
Vet. fees and sundries	3		5–6
	38		75–89

The quantities of concentrates used are relatively small but in percentage terms the costs that may be saved by home-mixing are substantial due to the relatively high price of proprietary calf foods.

BEEF CATTLE

Having reared the calf to three to six months of age we now have to decide whether it is worth keeping any longer. To make this decision we need to assess the margin we can expect from keeping the cattle and compare this to the opportunity cost of this decision to keep them. In short, we need to consider what else we could do with the land and capital that would be used to keep the cattle.

A six- to eight-month-old calf should weigh in the region of 150–200 kg and be worth £165–£220 (£1.10 per kg) if sold in April–May. The costs incurred in keeping it for a further six months are relatively small—say £10 for grazing variable costs, £12 for concentrate feed, and £15 for interest on capital. But the price of beef cattle falls to a low in the autumn and the outcome could be as shown in Table 7.5.

The effective price received for the additional liveweight gain is only 68p per kg due to the fall in overall price per kg. The extent of this fall is always difficult to predict as it varies considerably from year to year according to trends in beef supplies relative to

Table 7.5. Beef cattle margins—summer

		£
Value @ 12–14 months	300 kg @ £0.96	288.00
Value @ 6–8 months	200 kg @ £1.10	220.00
Output	100 kg @ £0.68	68.00
Less		
Feed	100 kg @ 12p	12.00
Grazing	0·14 ha @ £75.00 per ha	10.50
Interest on capital	(£220* for half-year @ 14 per cent)	15.40
		37.90
MARGIN TO COVER FIXED COSTS		30.10

* To be absolutely correct we should also charge interest on the cost of the food and grazing as well as the sale value of the animal.

demand and to the availability of winter feed supplies. The fall in most years, however, is substantial and the general rule would be to sell the cattle before the autumn unless one can be sure of having adequate winter feed supplies to carry them through to the following spring.

Table 7.6. Beef cattle margins—winter

	Store system		*Finishing system*	
Value @ 18–20 months	400 kg @ £1.02 =	408	450 kg @ £1.00 =	450
Value @ 12–14 months	300 kg @ 96p =	288	300 kg @ 96p =	288
	100 kg @ £1.20	120	150 kg @ £1.06	162
Less				
Variable costs				
Feed	350 kg @ 12p	42	600 kg @ 12p	72
Sundries		8		8
Forage costs		—*		—*
Interest on capital	£288 @ 14 per cent for half-year	20	£288 @ 14 per cent for half-year	20
		70		100
MARGIN TO COVER FIXED COSTS AND FORAGE COSTS		50		62

* See text for explanation
† To be absolutely correct one should also charge interest on the cost of the food and grazing as well as on the sale value of the animal

The output and costs incurred during the winter period depend on whether the animal is kept on a store or finishing ration. In turn, the latter depends to a large extent on whether they are inwintered or outwintered. The results could be as shown in Table 7.6.

The forage costs have purposely not been estimated because on some farms this will take the form of silage; on others it may be hay or straw. If it is the latter, these have a sale value and it would be appropriate to enter the sale value to determine whether or not this enterprise was worthwhile.

If the forage took the form of hay the amount required would be in the region of one tonne and its value would be roughly equal to the margin calculated above. This is what one normally expects to find with a beef cattle enterprise, *i.e.*, very little or no margin in the winter months if account is taken of *both* the finance charges and opportunity cost of all foods fed. This does not mean that over-wintering beef cattle can not be justified on a farm. If surplus

Table 7.7. Forecast results for autumn-born beef calves

	Store	£	Finished	£
Value at 20 months	400 kg @ £1.02	408	450 kg @ £1.00	450
Value as a calf (inc. mortality)	40 kg @ £1.50	60	40 kg @ £1.50	60
Output	360 kg @ 96·6p	348	410 kg @ 95·1p	390
Variable Costs				
Feed: 0–8 months	360 kg	60	360 kg	60
8–14 months	100 kg	12	100 kg	12
14–20 months	350 kg	42	600 kg	72
	810 kg	114	1,060 kg	144
Sundries		14		14
		128		158
Gross margin before forage costs		220		232
Forage costs—grazing	0·10 hectare			
—conservation	0·15 hectare			
	0·25 @ £100	25		25
GROSS MARGIN PER ANIMAL REARED		195		207
GROSS MARGIN PER HEAD PER ANNUM		117		124

silage is available and capital is also available free of interest then the margin shown above of £70 to £82 per head before finance charges is well worth having in relation to the labour and other fixed costs that would be involved.

The beef cattle enterprise has purposely been broken down into its component parts, *i.e.*, calf rearing, summer grazing, and winter finishing and store periods to illustrate the main profitability points and to show how with such an enterprise it is necessary to treat finance charges as part of variable costs.

When these component parts are combined the results that may be expected for an *autumn*-born calf are outlined in Table 7.7. It needs to be stressed, however, that beef is like any other farm enterprise in that very wide variations can be expected in profitability between farms. So far as an individual farmer is concerned the main point he has to establish is the optimum time of sale relative to the *'spare'* resources, including food, he has available.

The above figures show margins that are well above those actually achieved in the past two to three years—in 1978 to 1980—as the current (1981) prices of beef bears a favourable relationship to the current cost of cereals.

This relationship is fundamental to the profitability of beef production. The effect of a ten per cent increase in feed costs coupled with a ten per cent reduction in beef prices is illustrated in Table 7.8.

Table 7.8. The importance of the beef/cereal price ratio

	Store system £	Beef system £
Margin shown in Table 7.7	195	207
Less 10 per cent sale price	41	45
	154	162
Less 10 per cent increase in feed costs	11	16
REVISED MARGIN PER ANIMAL REARED	143	146

SHEEP

A breeding ewe enterprise is difficult to justify on the vast majority of dairy farms because it is very competitive for grazing at a critical time in dairy farming, the spring,

Breeding ewes produce a low gross margin per hectare and

cannot compete with cereals on an area basis. They tend to be less profitable than beef so are really only justified where there is a large area of grassland that is not suitable for dairy cows and where there are inadequate buildings to keep dairy replacements and beef cattle.

The fencing requirements for sheep are also extremely critical and this limits their introduction in a form that is beneficial, *e.g.*, as a store lamb fattening enterprise. The grazing of pastures by sheep in the autumn produces a considerable benefit but this coincides with a very busy time on dairy farms, particularly with autumn-calving herds. Fencing permitted, therefore, one of the best ways to gain this benefit to the pasture is to provide autumn and winter 'keep' facilities for sheep farmers on condition that they are responsible for shepherding.

NON LAND USING ENTERPRISES

Many dairy farms are relatively small and family farms may not be large enough to provide a 'living' for all the family without the introduction of a non land using enterprise.

Calf rearing is such an enterprise and the possible returns from it have already been discussed. 'Barley beef' is a non land using enterprise but its justification on a dairy farm is most unlikely.

Pigs and poultry are possible enterprises. Which one is introduced will depend on the interests of the family or farmer concerned. Egg production is now very specialised and its introduction is unlikely. The introduction of a small table poultry or turkey enterprise concentrating on the Christmas market can and does offer a very useful contribution to profits. However, a pig enterprise can be very complementary to the dairy operations, particularly as a means of saving fertiliser costs. These costs are in the region of £50–70 per cow and most of these can be saved on an intensive pigs and dairy farm by making effective use of the pig manure. On an all-grass farm there would also be potential benefits to be gained from sharing the machinery costs and/or organic irrigation costs of both cow and pig manure handling systems.

NON FARM ENTERPRISES

This chapter is concluded by a brief mention of the need to consider such non farm operations as providing tourist facilities, whether it be bed and breakfast or a caravan site.

As with non land enterprises, the objective here is to make a profit from resources which would otherwise be idle such as an ex-farmworker's cottage or part of the large farmhouse. In other cases the object will be to expand the business—by introducing a retail milk round or a milk processing enterprise, for instance.

Chapter 8

FARM BUSINESS ANALYSIS AND PLANNING TECHNIQUES AND THE DAIRY FARMER

COMPARATIVE ANALYSIS

A brief history of the development of comparative analysis techniques was given at the beginning of Chapter 3, the basis of comparative analysis being 'What one farmer can do another should be able to do too!'

There are several limitations to the value of comparative analysis as a method of improving farming profitability. The most important is that it is historical, and in periods of inflation the information provided is out of date before it is printed. Nonetheless it provides a valuable analysis tool for the farm management expert called in to advise a farmer on the future development of his business. By comparing the farmer's results to standards an adviser can quickly identify weaknesses in the system and in enterprise efficiency. Comparative analysis, however, only tends to tell what is wrong with a business; it does not necessarily tell what needs to be done to put it right, nor does it always show why it is wrong.

Having identified the weaknesses and strengths in a business, the next step is to identify the reasons for these weaknesses and then take appropriate steps to put them right. These steps may necessitate a complete replanning of the business organisation and will involve budgeting and other management techniques.

A second problem with comparative analysis is the wide range of performance concealed by the so-called average data. This is rectified to some extent by showing separate results for, say, the top 25 per cent or bottom 25 per cent of the sample. Often, however, it has to be accepted that there is no such thing as a typical average farmer. This is demonstrated, for example, by the data on milk yields shown in Table 5.2. In the analysis of results of dairy farms according to yield, 28·4 per cent of the farms have a yield between 5,000 and 5,499 litres; 21·2 per cent a yield between 5,500 and 5,999 litres; 21·0 per cent between 4,500 and 4,999; a further 29·4 per cent have a yield either above or below these levels. Relatively little is to be gained, therefore, in comparing your results to the average.

Considerable caution also has to be exercised when drawing conclusions from a table such as the one referred to above. In this particular table the farms are selected according to yield and show a significant increase in yield for a modest increase in concentrate costs. A protagonist of the economic benefits of feeding concentrates to improve yields would use this table as evidence to support his belief. If, however, the same farms are selected according to the 'amount of concentrates fed' a different grouping would emerge and this would show a very modest increase in yield for a given increase in concentrates. A protagonist of the 'need to cut concentrates cost school' would use this table to support his belief! To overcome these objections a third person might select according to the 'margin per cow'. He too would get a different grouping with a 'result' of concentrate increment: yield increment midway between the other two.

Most financial survey results are presented in terms of results per hectare; farms are classified as being above average or below average according to their result per hectare, and this is defined as Net Farm Income or Management and Investment Income. An above-average level is then assumed to equate with 'above average' management ability and efficiency. This to a large extent is true but what is not properly taken into account is the considerable additional capital per hectare or per acre that is associated with the good result. The additional capital invested is responsible for a significant proportion of the additional profit and there is need for a new definition of profit to take this into account. Alternatively much more attention has to be paid to return per £100 capital.

This need to look carefully at capital investment per hectare is particularly true in dairy farming where high profits per hectare tend to be associated with higher stocking densities and therefore higher capital investments per hectare. Care in this connection has to be taken to compare farms of a similar size because in general the smaller the dairy farm the greater the intensity and the greater the profit per hectare.

To sum up, therefore, considerable care has to be taken when comparing your results to 'standards'. Remember that the best standards are the results you achieved on your farm last year. More is to be gained by studying in detail the differences in results of a few farms you know well than by studying highly technical and computerised results from a large number of farms following different systems and farming in different environments. Beware in particular dairy enterprise results quoted from data not backed up by full farm costings.

One final point before leaving comparative analysis. The gross margin data is usually fairly reliable and not dependent on many subjective judgements other than stock valuations. The fixed costs on the other hand are very dependent on subjective judgements and much less reliable. Labour costs on small dairy farms are not assessed very accurately due to the high proportion of labour supplied by the farmer. Power and machinery costs too can sometimes be more dependent on the number of sons and hence car and domestic fuel bills incurred than on the actual costs of running the farm. Rent charges are also arbitrary on owner-occupied farms and are seldom worked out in relation to the actual quality of the land and buildings available.

Despite these reservations some comparative standards are essential if you are to assess your management capability and competence as a dairy farm manager.

THE GROSS MARGIN SYSTEM

The gross margin system as such had not been invented when the writers were at college and teaching farm management in the late 1950s and early 1960s. Its development in and since the 1960s has had much to do with the progress of farm business management. Virtually everyone in farming now talks about and uses gross margins and assesses enterprise performance in financial as well as in physical and husbandry terms. Criticism is levied at, and attention is drawn in this book to the limitations of, the term 'Margin over Concentrates' but twenty years ago very few people had heard of it.

The main advantage of the gross margin analysis and costing system is its simplicity and the ease with which data can be collected and used for planning purposes. It is, however, much more suited to the needs of the arable farmer than to the needs of the dairy farmer.

The gross margin system is of most value when considering minor changes in the cropping and stocking of a farm, as for example whether to grow four hectares of barley instead of four hectares of wheat, or keep eight more cows. Such changes do not lead to any consequent changes in fixed costs so the net effect on the profit is simply the change in the gross margin—but is it? In the case of the choice between wheat and barley there is no problem. The change is equal to the difference in the gross margin because such a change is not likely to affect any of the fixed costs *but* in the case of the additional cows there would also be a consequent change in finance charges. This consequent change in finance charges is often forgotten because they are placed in fixed costs 'for convenience'. Finance charges *do* change in proportion to a small change in the size of an enterprise but the problem is that they *cannot* be easily allocated.

The choice between eight more cows or four more hectares of cereals should be budgeted as shown in Table 8.1.

Table 8.1. Comparing gross margins and finance charges to choose between enterprises

	£
Gross margin from 8 cows @ £350 per cow	2,800
Less	
Finance charges to fund 8 additional cows	
= £4,000 @ 15 per cent	600
Net contribution from 8 additional cows (A)	2,200
Gross margin from 4 hectares wheat @ £450 per hectare	1,800
Less	
Finance charges to fund 4 hectares	
= £1,600 @ 15 per cent	240
Net contribution from 4 hectares wheat (B)	1,560
Total A − B = increase in profit by keeping 8 cows instead of growing 4 hectares wheat	640

There is also a tendency to assume that other fixed costs will stay the same for *substantial* changes in the size of individual enterprises, *e.g.*, an increase in cow numbers from seventy to one hundred. This may be true on a farm that is not using its buildings, labour *and* machinery to capacity but on most farms it is only partly true. Most farmers making such a change would have to incur some additional expenditure on buildings and equipment, with a consequent change in the mortgage and/or rent and an increase in machinery depreciation. The labour bill and other overhead costs would, however, probably not change.

Mistaken conclusions are often drawn when analysing accounts prepared on the gross margin/fixed cost basis. An example of this would be on a farm where the fixed costs per hectare were £420 and the gross margins were £650 for dairy cows and £280 for dairy followers.

An understandable reaction would be that the dairy followers were not covering the allocation of fixed costs per hectare. In fact, the situation could be that the dairy followers were run on rough grazing land which was completely inaccessible and unsuitable for the dairy herd, and if not used by followers would have no alternative use.

Another category of stock which has suffered as a result of the gross margin system has been the sheep enterprise. The sheep gross margin is quite low and compares unfavourably with dairy cows. Once again, similar to dairy followers, sheep can utilise marginal grazings unsuitable for the dairy herd. Additionally there is a bonus difficult to measure resulting from the use of a sheep flock to take off all the surplus autumn grass and prevent the possibility of winter kill in short-term leys. Sheep will also help to establish maiden seeds prior to grazing by the dairy herd.

In this connection the reader is reminded that the gross margin system is simply a means of trying to implement some of the economic principles outlined in Chapter 2. Unfortunately the system is so simple that most people forget the complementary and supplementary relationship between enterprises such as those discussed in Chapter 2 and Chapter 7. They also forget that these fixed costs do change if one makes substantial changes in the cropping or increases the intensity of stocking. This may not happen in the short term but in the long term there is a tendency for fixed costs to catch up with the increased intensity, *e.g.*, a bigger forage harvester is purchased, say, two years after expansion because it is *justified* by the expansion in herd numbers. Two years earlier one of the main reasons for justifying expense was the opportunity it would present to 'spread machinery costs!'

PARTIAL BUDGETING

Prior to the introduction of the gross margin systems, the principle of fixed and variable costs was implemented by means of partial budgets. A partial budget is set out in Table 8.2 which illustrates the same example as shown in Table 8.1, that is the replacement of four hectares of wheat by eight dairy cows.

The main advantage of the partial budget is that one considers each cost item in turn and asks 'In what way will this cost item change as a result of the change in cropping and stocking?' The main disadvantage is that it is a cumbersome and time-consuming way of considering several alternatives, *e.g.*, barley compared with cows, compared with wheat, compared with dairy replacements.

Table 8.2. Partial budget to assess effect of replacement of 4 hectares wheat by 8 dairy cows

Extra costs	£	Costs Saved	£
Replacement heifers 2 @ £550	1,100	Wheat seed 4 @ £35	140
Concentrate feed 8 @ £250	2,000	Wheat fertilisers 4 @ £75	300
Veterinary & sundry costs 8		Wheat sprays 4 @ £65	260
@ £35	280	Wheat sundries 4 @ £30	120
Forage costs 4 ha @ £150	600	Labour	Nil
Labour costs	Nil	Power and Machinery	Nil
Power and machinery costs	Nil	Finance charges 4 @ £400 @	
Finance charges 8 @ £500 @		15 per cent	240
15 per cent	600		
Straw 10 tonne @ £20	200	*Revenue Gained*	
		Milk sales 8 @ £700	5,600
Revenue Foregone		Calf sales ·8 @ £55	440
24 tonne wheat @ £100	2,400	Cull cows 2 @ £360	720
	7,180		
Net gain from change	640		
	7,820		7,820

One of the problems with partial budgeting as with all planning is the need to make decisions without really having the knowledge or time to work out the effect of the proposed change on the profit. For example, you should pose the question 'Will it pay me to purchase my own equipment to make silage rather than use a contractor?'

A partial budget to help answer this question would need to be set out as shown below:

Extra Costs	*Costs Saved*
Machinery depreciation	Contract charges
Interest on additional capital invested in machinery	Concentrates
Fuel and repairs	
Labour	
Revenue Foregone	*Revenue Gained*
	Milk

The extra costs would be those associated with the purchase and use of the silage-making machinery, including machinery depreciation, fuel and repairs, and the additional labour costs that might be incurred. Against these one can offset the savings in contract charges and the expected benefits 'better silage' would have on the Milk Sales and Concentrate Costs. The latter are the most difficult items to evaluate.

Having worked out the above, one would then have to consider the question 'Would it pay better to invest the capital in more cows and/or another enterprise?'

COMPLETE ENTERPRISE COSTINGS

Prior to the introduction of the gross margin system it was normal practice to carry out complete enterprise costsings which involved allocation of fixed costs to individual enterprises as well as variable costs.

The time taken to carry out full enterprise costings is much longer than for gross margins and much of the information produced is of little value. However, on the larger dairy/mixed cropping and dairy farm there is a growing need for 'fixed costs standards' for individual enterprises as well as fixed costs standards for farming systems.

Farmers are exhorted to watch their fixed costs but at the moment there is little information available to indicate what the fixed costs should be at the individual enterprise level. Most herd managers, for example, can readily quote their 'Margin over Concentrates', 'Stocking Rates', 'Calving Index' *etc.*; all figures related to the gross margin. Very few, however, have at their finger tip the 'Cost of Labour per Cow', or their 'Power and Machinery Costs' per hectare of forage. One of the reasons is the time-consuming nature of the records that are required but this is one way in which computers may come to be used more extensively in the future, *i.e.*, to investigate costs that cannot be determined at the present time due to the expense of recording this information. It is stressed, however, that this is only likely to be necessary on the larger and the mixed dairy/arable farm. The specialist small dairy farmer will continue to find his time and efforts best devoted to the existing gross margin/fixed cost system.

GROSS MARGINS AS A MEASURE OF GRASSLAND EFFICIENCY

In dairy farming the gross margin is used to measure two things at the same time, *i.e.*, the efficiency of dairy cow management and the efficiency of grassland management.

Let us just look for a moment at an arable farmer who keeps pigs and grows cereals to feed to these pigs. He works out a gross margin from growing cereals and a separate gross margin for keeping pigs, *but* if he did not keep a record of barley yields he would have to be content with a pig gross margin per hectare. The latter would be of very little use to him as a low figure could be due either to a poor performance by the crops or a poor performance by the pigs. He therefore keeps records of yields so that the two can be separated.

We have a similar problem on the dairy farm *but* because it has always been accepted that we do so, we try to assess the performance of both the cows and the grassland with one figure. At the end we do not really know whether it is the cows that are good or the grassland management that is good, or poor, as the case may be.

If we take an extreme case—the place of beef cattle on the farm—it is generally accepted that these produce a low gross margin per hectare. If, however, we take calf rearing and express this on a per hectare basis we find it is very high; but this does not mean that calf rearing is a good way of utilising grassland. It is simply due to the fact that they use relatively little land relative to other feed requirements.

We do not keep records of the actual grazing and conservation requirements of dairy cows on the one hand and young stock on the other. By 'we' in this case is meant not only farmers but also all the many advisers and commercial organisations who collect financial and other information about dairy farms. We adopt the livestock unit basis to allocate the forage costs and finish up with a mass of misleading information.

The problem is not easy to solve because of the problems involved in grassland recording. It would seem, however, that more could and should be done to separate the assessment of grassland production efficiency from dairy cow and other grazing livestock efficiency. This need is particularly acute in trying to resolve the conflicting views on the use of concentrates as supplements to silage or grazing.

How can a start be made to solve this problem? It is suggested that this can only be done by attempting to evaluate separately the financial contribution on the one hand of grazing and silage, and on the other the financial contribution of the dairy cow. To do this the grazing and conserved products need to be charged to the dairy cows at their opportunity cost or sale value. It is accepted that silage and grazing do not often have real sale values but these can be computed based on hay equivalents and known sale values of grazings.

If this is done the result could be along the lines shown in Table 8.3.

Table 8.3. Separate assessment of grassland and dairy cow contributions to the total Dairy Herd Gross Margin

(i) Grassland Contribution	Grazing £	Silage £	Total per hectare £
Value of grazing 400 cow days @ 60p	240	—	240
Value of silage 14 tonne @ £20	—	280	280
	240	280	520
Variable Costs			
Seed	5	5	10
Fertilisers	60	90	150
Sprays	5	5	10
Sundries, inc. silage preservative	10	20	30
	80	120	200
GRASSLAND CONTRIBUTION	160	160	320

(ii) Dairy Cow Contribution		Per litre p	Per cow £
Milk sales	6,000 litre @	13.5	810
Less concentrates and other purchased feed		4.5	270
Margin over concentrates and purchased feed		9.0	540
less Grassland at Opportunity Cost (0.50 hectares)			
Grazing 200 days @ 60p		2.0	120
Silage 7 tonne @ £20		2.3	140
Margin over all feed at opportunity costs		4.7	280
Add calf output			60
Subtract herd depreciation			(50)
Subtract veterinary and sundry costs			(40)
DAIRY COW CONTRIBUTION			250

(*iii*) *Total Gross Margin*   *£ per hectare*
Dairy Cow Contribution £250 × 2   500
Grassland Contribution   320

TOTAL GROSS MARGIN   820

(*iv*) *Grassland Contributions to Profit*	*Grazing £ per hectare*	*Silage £ per hectare*	*Total £ per hectare*
Grassland Contribution to Gross Margin	170	160	320
Less Fixed Costs:			
Rent	40	40	80
Power and machinery	20	60	80
Labour	15	45	60
Sundry overheads	10	10	20
	85	155	240
CONTRIBUTION TO PROFIT	75	5	50

The data in Table 8.3 shows a Grassland Contribution of £320 and a Dairy Cow Contribution of £500 to the total Gross Margin of £820 per hectare.

The data also includes an indication of how the grassland fixed costs which are assumed to total £240 per hectare might be divided between grazing on the one hand and silage on the other. It is interesting to note that the silage output of £280 per hectare, 14 tonne at £20 per tonne is only £5 more than the assumed fixed costs.

Table 8.4. Margin over all feed at opportunity costs

	£
Milk sales; 6,000 litres at 13·5p	810
less	
Concentrates 2·0 tonne @ £135	270
Margin over concentrates	540
Less	
Grazing costs or 200 days @ 4.125p/day	82.5
Silage costs 7 tonne @ £19.64 p/tonne or 165 days @ 8.333 p/day	137.5
	220
Margin over all feed at opportunity costs	320

This point takes us back to the comment made earlier about the need for more complete enterprise costings so that data can be produced on the lines shown in Table 8.4.

It is accepted that opportunity costings are more time consuming than conventional gross margin costings but costing services of this nature need to be developed if answers are to be obtained to the economic problems in assessing grassland productivity.

BUDGETARY CONTROL AND MANAGEMENT BY OBJECTIVES

In Chapter 1 we highlighted the two basic functions of management, namely deciding what to do and doing it.

The efficient dairy herd or dairy farm manager does not simply compare his results to standards. He prepares a physical and financial target for the year, month, or week and then compares his result to the target. This applies whether he is preparing the budget for the whole farm or comparing his level of intended feeding of the cows to what actually happened.

If he is going to do this he needs to have the information available quickly and in a form that is readily understood. Such a system is described later in the book in Chapters 10 and 11.

The basic objective of the recording and accounting systems described is to have information available in a form and at the time it is needed to prevent things going wrong, or 'to allow opportunity to be taken of favourable chances'.

Section 2

Setting Up and Managing a Dairy Farm

Chapter 9

RESOURCE ASSESSMENT

In this section it is assumed that you are taking over the management of an 84 hectare dairy farm. The objective is to deal with the main points you should consider in arriving at your management strategy, and the general approach will be much the same if you are taking over as a tenant.

ASSESSING THE FARM AND ITS RESOURCES

When taking over a farm you need to guard against imposing a preconceived plan on it. It is most important that you assess all the farm resources carefully and then prepare a plan to suit.

(a) *The Land.* Examine the nature of the land and try to assess the inherent fertility and the suitability of each field for grazing, conservation or other crops. In particular assess whether there is a *substantial* area that cannot be grazed by the dairy herd; this will indicate whether there is a basic need for a youngstock enterprise. You should also assess the difference between the total *effective* hectares and the total farm hectares including roads, buildings and woodlands, *etc.*

Given an outcome such as that shown in Fig. 5 it would *not* be prudent to plan for more than 120–130 cows because the total area that can be grazed by the cows is only 52 hectares. Youngstock and or cereals would need to be grown on the remaining area.

In addition, you need to take careful note of which fields require drainage and/or other improvements. Drainage and pasture quality may limit cow numbers to 100 despite there being 52 hectares available. It is important to try to assess whether poor swards are due to poor management or poor inherent fertility.

(b) *Water Supplies, Roads and Fences.* These must also be taken into account when assessing the current capability of the land and its future potential. A map of the farm, scale 6 inches to the mile or 1 in 10,000 is an invaluable aid in preparing the cropping plan.

| Field number | Hectares | Suitable for: | | | | | General remarks |
		Cow grazing but not conservation	Cow grazing or conservation	Youngstock grazing or conservation	Youngstock grazing only	Cereals	
1	6	6	—	—	—	—	Badly poached
2	8	8	—	—	—	—	Good ley
3	4	4	—	—	—	—	Needs drainage
4	8	—	8	—	—	—	Water supply limited
5	6	—	6	—	—	—	ditto
6	4	—	4	—	—	4	
7	8	—	8	—	—	8	Poor ley
8	8	—	8	—	—	8	
9	8	—	—	8	—	8	Poor access
10	4	—	—	4	—	4	ditto
11	6	—	—	6	—	6	ditto
12	8	—	—	—	8	—	Unproductive
13	2	—	—	—	2	—	
	80*	18	34	18	10	38	

* Plus 4 hectares buildings, roads and waste areas

Assessing the land

(c) *Buildings and Fixed Equipment.* Assessment of these needs to deal with the following:

- Present use, potential use with minor capital improvements, potential use with major capital improvements.
- Number of cows, youngstock and calves which can be accommodated in the buildings.
- Ease or difficulty of feeding and managing the stock.
- Amount of fodder than can be stored in the buildings as silage, hay, straw or grain.

You should make a sketch plan of all the buildings on which materials and cow-flow diagrams can be superimposed as part of the assessment.

(d) *Farm Cottages(s).* The availability or otherwise of these will influence the choice of staff.

As a result of the above resource assessment you should be able to formulate a short-term cropping and stocking programme appropriate to the existing farm resources. From this a longer-term cropping and stocking programme can be derived given (i) minor capital and husbandry improvements, and (ii) major capital and husbandry improvements.

(e) *Farm Staff.* On a farm of this size there will be one or two staff already employed on a full-time or regular part-time basis. Whether you should continue with these staff may be a major policy decision; however, if you are appointed as manager it is likely that your employer will wish you to make effective use of the men already on the farm. This may well necessitate revising the policy you prepared based on your initial assessment of the farm.

(f) *Machinery and Equipment.* You should prepare an inventory of the machinery and equipment already on the farm. Note its state of repair, assess its value and its suitability to the farming system you have in mind.

Estimate the capital cost of the additional machines to be purchased together with the cost of those in need of replacement. Calculate the depreciation charge resulting from these proposed transactions, along the lines of the example in Table 9.1.

If you are taking over the farm as a tenant the list will be that of machines you already own. Your capital will be limited and the list of proposed investments will need careful scrutiny to ensure that the investment in machinery is justified in preference to investment in stock or other resources.

(g) *Livestock.* The procedure you should follow to assess your stock is as follows:

- Determine existing numbers and judge their condition.
- Judge their production potential (bearing in mind any problems posed by the absence of detailed records and accounts).
- Prepare a simple inventory and establish the value of the stock.
- Compare this inventory and valuation with that required for your policy.

(h) *Crop Produce and Stores on Hand/Tenant Right.* The quantities on hand will depend on the time of year.

As the prospective manager you will be mainly interested in determing whether or not food supplies are adequate and in checking whether seeds and fertilisers for the coming year's crops are on hand or not.

If you are a prospective tenant this area of assessment is vitally

Table 9.1. Machinery inventory

Machine	Notes	Present value (A) £	Proposed net outgoing (B) £	Total (A+B) £	Budget depreciation £	Estimated closing valuation £
Tractor		5,000		5,000	1,250	3,750
Tractor	Needs replacing	2,000	4,000	6,000	1,500	4,500
Tractor		1,000		1,000	250	750
Forage harvester	Poor maintenance	2,000		2,000	400	1,600
Silage trailers (2)		2,000		2,000	400	1,600
Mower	Needs replacing	200	600	800	160	640
Heavy roller		600		600	60	540
Buckrake		200		200	20	180
Fertiliser spreader	Scrap value		600	600	90	510
Plough		400		400	40	360
Cultivating equipment		400		400	40	360
Sprayer	Scrap value	—		—	—	—
Slurry tanker		400		400	80	320
FYM spreaders (2)	Poor condition	1,000		1,000	250	750
Farm vehicle	Poor tyres	3,000	—	3,000	750	2,250
Haymaking equipment		Nil		—	—	—
Baler		—		—	—	—
Milking equipment		4,000	—	4,000	400	3,600
Bulk tank		2,000	—	2,000	200	1,800
Contingency			800	800	200	600
TOTAL		24,200	6,000	30,200	6,090	24,110
Per hectare*		302·5	75	377·5	76·1	301·4

* Effective area. Roads and buildings not counted

important because it will have a considerable effect on the capital required to take over the farm.

As an incoming tenant you will be expected to pay the market value for all crop produce and stores on hand. In addition payments may well be necessary in respect of 'unused manurial values' and tenant's fixtures. This last point takes us back to the assessment made of the water supplies, fences, hedges, buildings and other fixed equipment. As a prospective tenant you will have a keen interest in determing the likely capital sum you will need to pay the outgoing tenant for his 'Tenant's Improvements and Fixtures'. In addition you will have made enquiries and taken note of any dilapidations that may be charged by the landlord or that you may be taking over from the existing tenant.

You will also need to look again at the buildings as the landlord may be prepared to carry out improvements provided you are also prepared to pay the appropriate increase in rent to fund these improvements.

ANALYSIS OF PREVIOUS YEAR'S FINANCIAL PERFORMANCE

Taking over as manager you should have access to the previous year's financial results. These need to be examined with these particular points in mind:

(a) *What are the likely levels of fixed costs?* This information is going to be vital in the next stage of the venture—preparing your first year's programme together with the annual gross margin fixed budget.

(b) *What level of grass margin can you expect from the individual livestock enterprises?* Are there major weaknesses in enterprise management? If so can they be corrected quickly, or do they necessitate below-average results in the future until such time as the inherent problems leading to them can be overcome?

(c) *Are there basic weaknesses in the farming system such as too many youngstock or too few cows?* If so, is this simply due to a poor understanding of the profitability factors in dairy farming? Or is it due to lack or resources in the form of working capital to purchase the stock or fixed capital needed to provide additional buildings and fixed equipment?

The basic objective of the analysis is to gather information that will allow you to do the following:

- Prepare a financial budget for your first year of operation that is likely to be fairly accurate.
- Quickly identify weak areas in enterprise management that can be corrected before they occur again.
- Decide on a farming programme that is appropriate to the needs of the business.

ASSESSMENT OF OVERALL CAPITAL POSITION

It is essential to make this assessment *before* moving on to the detailed planning of the cropping and stocking of the farm, and the preparation of the whole farm budget for this detailed plan. If not you could well find you have prepared a detailed plan that is completely outside your own financial resources as a prospective tenant, or the financial resources of your employer if you are a manager.

In the case of a farm manager the first stage in this assessment is to detail and evaluate the farming assets already on the farm, and at the same time to list the liabilities in order to arrive at the net capital situation (see illustration in Schedule 1, Appendix 3).

The second stage is to consider ways in which the allocation of capital resources could be changed to bring about a better liquidity position and/or potentially more profitable farming system.

The third stage in this capital assessment is the production of an annual cash-flow budget to ascertain how the asset requirement/borrowed capital requirement changes during the year. How this is done is illustrated in the next chapter.

The approach of a prospective tenant farmer is very similar. In this case he lists the resources he proposes to employ on the farm to find the total initial capital requirement. In his case he must not forget the capital required to meet the tenant's ingoing.

RECURRING NATURE OF RESOURCE ASSESSMENT

This chapter has emphasised the need for resource assessment when taking over a new farm. However, resource assessment is not simply a one-off study; it is a policy that should continue throughout your period of management. As time progresses the assessment becomes more detailed and more accurate because it is based on experience.

Chapter 10

PREPARING THE ANNUAL BUDGET AND DETAILING MANAGEMENT OBJECTIVES— A CASE STUDY

Your detailed assessment of the farm and its resources will have given you a good idea of the cropping and stocking policy you would like to implement. You will also have lots of ideas involving capital expenditure that you would like to put into operation.

If you are a manager you should have access to the previous year's financial results and these will be used as a base from which to construct the budgets. By the time you reach the end of your first full financial year you will have the results of your own efforts as well as your budget. This is the real test of management: comparison of actual results to budget.

To illustrate this process of budgetary control it is intended to use a case study on an actual farm, and data relating to the farm is shown in Appendix 3. The data has of course been modified where necessary to maintain confidentiality.

1. RESOURCE ASSESSMENT

(a) *Land.* 24 hectares of the 80 effective hectares are in permanent pasture. These are not suitable for conservation and 16 hectares are difficult to graze with in-milk cows. The remaining 56 hectares are well suited to intensive grassland production.

(b) *Buildings and Fixed Equipment.* Cubicle housing is available for 140 cows and there is an open grass silo suitable for self-feeding. Supplementary feeding of concentrates outside the parlour is also feasible and this allows the manager to aim for high yields.

The slurry storage facilities are not ideal and the milking parlour is somewhat out of date by modern standards. The manager would like to see additional capital expenditure incurred to improve the parlour and to improve slurry storage.

(c) *Farm Staff.* The manager has an assistant. Whether or not a third person is employed depends on the manager's inclinations and whether he can produce results adequate to justify the employment of this third person and at the same time provide cash to fund capital improvements.

(d) *Machinery and Equipment.* This is adequate but not substantial. Silage making is done on a contractual basis and this has worked well in previous years.

(e) *Livestock.* Numbers at the start of the financial year are expected to be as follows:

Dairy cows (mainly autumn calving)	140
In-calf heifers (due Sept/Oct)	40
Heifers under 1 year	45

(f) *Crop Produce/Stores on Hand.* Silage supplies are adequate. There should be no need to purchase fodder during the first few months of the coming year. *Note:* The financial accounting year starts on 5th April.

Fertilisers are purchased in February and it is anticipated that all next year's crop will have been paid for before the start of the next financial year.

(g) *Previous Year's Performance.* This information is available and the current year's performance to date is also available. The Margin over Concentrates per cow in the year ended 31 March 1980 was £440. The margin for the 10 months ended 31 January 1981 is £396 and it is expected to be £500 for the 12 months ending March 1981, see Schedule Three.

Note: This chapter is being written assuming that the manager has been asked in February 1981 to produce budget for the year commencing 1 April 1981. This is a good time to prepare the gross margin and fixed costs budgets and to decide on the strategy for the coming year. It will be seen however as this chapter progresses that preparing budgets before the start of the year does present problems particularly from a cash-flow monitoring point of view. The *easiest* time to produce budgets is after the results for the previous year are known, that is in May/June for a year beginning 1 April, but this is often too late from a decision-taking point of view.

(h) *Capital Position.* For the purpose of this excercise it is assumed that the borrowed capital will total £30,000 at the start of the financial year and that the rent payable is £87.50 per hectare or £7,000 per annum.

An estimate of the capital invested in this dairy farm is given on Schedule One, Appendix 3. Farm assets excluding land and buildings total £140,500 or £1,756 per hectare. The investment net of borrowings and trade creditors is £98,000 or £1,225 per hectare.

A profit before finance charges of £19,600 or £245 per hectare is required to show a return on capital investment of 20 per cent.

Note: All the per hectare figures are based on the effective area of 80 hectares. To be strictly accurate one should also take into account the 4 hectares of buildings and roads but this leads to confusion in setting out the budget data.

2. PROPOSED FARMING PROGRAMME

(a) *Dairy Herd.* It is proposed to continue with a dairy herd averaging 140 cows and to continue to seek to achieve high yields and a high Margin over Concentrates per cow.

The emphasis is on autumn calving as can be seen from the Margin over Concentrates data compiled for the year ending 31 March 1981 (see Schedule Three). The detailed monitoring of the yields of monthly calving groups has shown that the August- and September-calving cows perform less well than those calving at a later date so a decision has been taken to start calving a month later in the coming year.

A case can be made for delaying calving until October but such a decision has been deferred due in part to the pressure this would impose on the rather modest calf-rearing facilities that are available at the present time.

(b) *Dairy Replacements.* The aim is to rear dairy heifers to calve at two years of age. This objective is being achieved because calves born in the autumn/winter of 1979–80 have been put to the bull and 42 are expected to be in-calf at 31 March 1981. It is anticipated that 42 heifers will be 10–14 more than the number required to replace cull cows. The surplus will be sold as down-calving/newly calving heifers.

(c) *Cropping and Forage Strategy.* The planned utilisation of the 80 hectares is as summarised in Table 10.1. It is expected that tactical changes will become necessary as the year progresses but no major revisions are anticipated.

In an attempt to contain fixed costs is is proposed to delay the first silage cut until the end of May and to aim to produce all the silage requirement for the dairy herd, *i.e.*, 840 tonne or 6 tonne per cow, in two cuts. A third cut producing 120 tonne is needed for the dairy replacements. A small amount of hay is fed to try to improve milk quality. It is planned to make 42 tonne, or 300 kg per head for the 140 dairy cows, and 18 tonne or 400 kg per head for up to 45 calves.

If the above is achieved it is anticipated that there will be a modest carry-over of silage stocks into the following year as past performance indicates a need for 5·0 tonne per cow. This is less than the normally accepted target of 7–8 tonne due to the high-yield strategy adopted on this farm. The achievement of carry-over stocks is an important management objective. It should be aided by the decision taken to put back the date of starting calving from August to September because all cows are brought indoors on to silage as soon as possible after calving.

Note: May 1981 proved to be one of the wettest on record so silage was not made until June, but yields were high as a result.

Table 10.1. The planned land utilisation

	April & May		June & July		August & September	
	No.	*Hectares*	*No.*	*Hectares*	*No.*	*Hectares*
Heifers <1 year	42	4 P.P.	42	4 P.P.	42	8 Ley
Heifers in-calf	42	8 P.P.	42	8 P.P.	} 80	24 P.P.
Dry Cows	—	—	30	6 P.P.		+ 6 ST/K
In-milk cows	140	12 P.P. + 16 ley	100	6 P.P. + 16 ley	80	30 Ley
	224	40	214	40	202	68
Silage		40		24		12
Hay		—		10		—
ST/K		—		6		—
		80		80		80

Note: P.P. = Permanent pasture
ST/K = Stubble turnips/kale

(d) *Staffing.* Three times per day milking is being introduced to improve yields in order to justify a full-time staff of three: manager, assistant herdsman and student.

(e) *Capital Expenditure Programme.* There is urgent need for a replacement tractor and farm vehicle. Major capital items are necessary in the long term but none are planned in the current year.

The budget capital expenditure based on second-hand replacements is as follows:

	£
Replacement tractor	3,500
Replacement farm vehicle	2,000
Contingency	1,500
	7,000 or £87.50 per hectare

3. GROSS MARGIN/FIXED COSTS BUDGETS

(a) Two Levels of Performance

The gross margin and fixed cost budgets for 1981–2 are detailed in Schedules 2, 4, 5 and 6 and are shown at two levels of performance, namely target and budget performance. The target performance represents the level of performance the manager hopes to achieve given good luck, good management and no abnormal factors such as drought. There is a possibility that these levels will not be reached and consequently decisions about capital investments are made in the light of the budget performance.

These estimates are prepared based on the prices and costs that may be expected in the 1981–2 financial year. The price : cost ratios in this particular year are expected to be less favourable than in the past and consequently it is necessary to assume high levels of technical performance in order to show a reasonable profit.

(b) Fixed Cost Estimates—see Schedule Two

(i) Labour costs are based on a labour force of three people and the detailed assumptions are shown in Table 10.2. *Note:* These are estimates of cash outgoings only. The true earnings of the staff would be greater than shown due to fringe benefits such as farm vehicles, houses and milk. It is of course unrealistic to assume that a farm manager would budget for his target salary to be lower than his budget salary but it is shown this way so that the possible result can also be seen from an employer's point of view.

(ii) The power and machinery costs are based on those incurred in previous years.

The proposed capital expenditure on machinery is assumed for simplicity to equal the depreciation charge, as detailed earlier in paragraph 2(e).

(iii) The rent, as previously mentioned, has been assumed to be £87.5 per hectare.

Property repairs vary considerably from year to year on a particular farm depending on whether or not major repairs are carried out in that year. Those shown are typical average figures.

(iv) Sundry overhead costs are influenced to a large extent by the need or otherwise for lime and overhead sprays, and the degree of insurance cover. Those shown are typical average figures.

(v) The overdraft requirement is assumed to be £30,000.

Table 10.2. Estimated labour costs

	Budget £		Target £
Manager's Salary, 160 per cent Grade I Worker = 160 per cent £86.40 = £138.24 × 52 weeks	7,189	(150 per cent)	6,740
Assistant's Salary, Craftsman Rate + 15 hours overtime = £73.60 + 15 × £2.76 = £115 × 52 weeks	5,980	(12 hrs O/T)	5,550
Student (aged 19), Weekly overtime 8 hours = £66.24 + 8 × £2.485 = £86.12 × 52 weeks	4,478		4,000
	17,647		16,290
add Employer's National Insurance contributions (14 per cent)	2,470	(12 per cent)	1,955
	20,117		18,245
Contingency including Increase in Wage Rates next January* say	883		755
	21,000		19,000

* *i.e.*, for last 2 months of financial year

Interest rates fluctuate widely from year to year. At the time of writing the base rate is 14 per cent giving an effective rate to most farms of 16·5 to 17·0 per cent. In the target performance it is assumed this falls significantly, *i.e.*, to 12 per cent, giving an annual charge of £7.20 per hectare instead of £10.00.

(vi) The total fixed costs at the budget level are £58,700 or £725 per hectare.

The scope for reduction in these costs is very limited and consequently the target fixed costs still total £53,400 or £667 per hectare.

(c) Gross Margin Estimates

Margin over concentrates. The factors determining this margin have been discussed at length earlier in the book. In this particular instance we have detailed information available for the year ended 31 March 1981 (Schedule 3). As mentioned before, a decision has been taken during the current year to commence milking three times per day and the manager's target projections for the coming year are shown on Schedule 4.

The manager is confident that the yield will continue to increase as the peak and current yields of the cows calving this winter are well above those of a year ago. The effect this has had to date on the monthly yield per cow in the herd is shown in Table 10.3 along with the manager's target for the coming 12 months.

Table 10.3. Monthly yield per cow

	1979–80 litres	1980–1 litres	1981–2 litres
April	505	593	750
May	532	614	670
June	406	457	540
July	341	317	400
August	342	243	300
September	346	364	360
October	379	514	520
November	462	576	580
December	576	735	740
January	663	831	840
February	666	755	760
March	682	831	840
TOTAL	5,804	6,800	7,300

He is also conscious that concentrate feed costs in 1980–1 have been somewhat higher than is really necessary due to the lack of adequate good grazing swards in the spring of 1980 and to a shortage of silage in the winter period. Steps have been taken to try to improve the grassland and, as already mentioned, a major management objective in the coming year is to increase the quantity of silage conserved. The attainment of these objectives should lead to savings in feed costs and these are allowed for in the target margin over concentrate estimates in Schedule 4. The target margin is £85,843, or 141·7 cows @ £606 per cow.

This represents a 20 per cent improvement on the previous year and in view of this the budget performance is reduced to £79,800, or 140 cows @ £570. Factors other than the MoC *do* affect the gross margin and to demonstrate the effect of those factors two estimates with the same budget margin over concentrate per cow are shown in Schedules 5 and 6. The margin over concentrates per cow in both these estimates is the same but the difference in the total gross margin is nearly £7,000. The reasons for this variance are summarised below:

- two additional cows kept throughout the year;
- five more heifers available for sale as down-calving heifers due to need for fewer herd replacements;
- improved sale price for dairy heifers and calves;
- modest savings in youngstock feed and in sundry costs.

There is relatively little difference between the target and budget forage costs as in this case most of this expenditure has been planned and incurred before the start of the year.

It should be noted, however, that the time of paying for these items can have a considerable influence on this cost and account for differences between farms. If the fertilisers were purchased and paid for after use, *i.e.*, in July/August instead of February, one could expect a 10 per cent increase in price. It should also be noted that a different planning decision would have made a significant difference to the expected fertiliser (see Table 10.4).

The extra cost of the alternative policy based on standard recommendations is £2,367 or 29 per cent more than the planned policy. Whether or not the planned policy is correct is debatable. However, it is planned to make as effective use as possible of slurry and fym to maintain soil P & K levels. The strategy may need to be revised in the longer term. A more liberal fertiliser policy can be more readily adopted in periods of relative profitability.

(d) Gross Margin/Fixed Costs Budget Summary

This is summarised on Schedule 2, and at the budget level it shows a potential profit margin of £16,900 or £211 per hectare, after allowing for £5,100 or £64 per hectare finance charges.

The budget profit before finance charges is £22,000 or £275 per hectare. The estimated capital investment is £140,500 or £1,756 per hectare giving a budget return on capital of 17 per cent.

The profit at the target performance levels is £34,048 and this

Table 10.4. Planning decisions and fertiliser costs

	Planning Decision		Alternative Decision	
(a) *Permanent Pasture* (24 hectares)	Use 160 kg N per hectare. Apply no P & K as no conservation and P & K levels satisfactory		Use 160 kg N and apply 26 kg P & K per hectare	
Budget Cost (per hectare):	15 bags 34·5 per cent N @ £5.00 =	£75.00	10 bags 34·5 per cent N @ £5.00 =	£50.00
			7·5 bags 22.11.11 @ £6.50	£48.75
			Per hectare	£98.75
		× 24		× 24
	Total		Total	
		£1,800.00		£2,370.00
(b) *Leys—Mainly Silage* (34 hectares)		£		£
First Cut:	7·5 bags 34·5 per cent N @ £5.00 =	37.50	7·5 bags 20·8·14. @ £6.00 =	75.00
Second Cut:	8·75 bags 25·0·16 @ £5.50 =	48.125	8·75 bags 20·0·16 @ £5.50 =	48.125
Grazing:	5 bags 34·5 per cent @ £5.00 =	25.00	5 bags 34·5 per cent N @ £5.00 =	25.00
		110.625		148.125
		× 34		× 34
		£3,761.00		£5,036.00
(c) *Leys—Mainly Grazing* 22 (hectares)		£		£
	22·5 bags 34·5 per cent N @ £5.00 =	112.50	17·5 bags 34.5 per cent N @ £5.00 =	87.50
		× 22	7·5 bags 22·11·11 @ £6.50·11	48.75
		2,475		136.25
				× 22
				£2,977
(d) TOTAL FERTILISER COST	£8,036		£10,403	
	or £100.45 per hectare		or £130.00 per hectare	

is nearly double the budget level. After deducting finance charges the target profit is £30,448 or 22 per cent of the capital invested.

These figures underline the problems facing dairy farmers at this time (1981–2) in their attempts to gain a good return on their capital. The budget performance levels are in themselves above average and in planning the farm business one must also take note of the possibility that the results may be below budget as well as above budget. The data at the foot of Schedule 2 shows what is needed to double the budget profit. They also show how easily the budget profit could disappear, as a similar adverse variance would give a loss of £248.

The average farmer in 1981–2 is likely to produce a gross margin per cow in the region of £375 and a gross margin per dairy replacement in the region of £90, *i.e.*, about 20 per cent below those detailed at budget levels for this farm. If the performance level on this farm was only average the budget for 1981–2 would be as shown in Table 10.5.

Table 10.5. Average performance figures for the example farm

	£	£ per hectare
Gross Margins		
Dairy cows 140 @ £375 =	52,500	
Dairy replacements 85 @ £90 =	7,650	
	60,150	752
Fixed Costs		
Labour	21,000	262
Power & machinery	18,000	225
Rent and property charges	11,000	138
Administration & sundries	3,600	45
Finance charges	5,100	64
	58,700	734
BUDGET PROFIT	1,450	18

Chapter 11

IMPLEMENTING THE FARM PLAN—MONITORING FINANCIAL PERFORMANCE

Having prepared your plans and objectives you now have to put them into operation. At the end of the year you have to be ready to see your results compared with the budget you made at the start. If you are wise you will set up a monitoring system that will allow you to correct adverse deviations from the plan as they occur. One of the main arts of management is to be aware of problems before they materialise and to have remedial actions ready to implement.

MONITORING MoC PERFORMANCE

A check can be kept monthly of financial progress compared to budget and to the results achieved in the previous year by recording the data as shown in Schedules 3 and 4 (Appendix 3). However, although this information is up to date it is *not* the only data you will need to ensure you are managing the herd to the best of your abilities. The monthly MoC data tells you how well you have done but if results are not as good as expected it does not tell you the reasons why, and it also tends to be too late for remedial action.

To answer this question you need to turn to the more detailed records you should be keeping of the progress of individual lactation groups, see Chapter 12.

A poor margin in a particular month will often reflect poor management at a much earlier date, such as lack of good-quality silage or a high proportion of dry cows due to problems with getting cows in-calf 10–12 months ago. We need therefore to be aware of these potential problems and to take action now that will lead to good results in the future.

The case-study data shown in Schedules 3 and 4 illustrates the effect of the right action now and in the recent past on the results in the future. Yields in January and February 1981 are up by 15–20 per cent compared to 1980 due to the adoption of three-times-a-day-milking and improved feeding practices. Consequently a similar increase can be anticipated during the next 4–6 months. It is also hoped to incur lower feed costs during the summer months as a consequence of pasture improvements carried out in 1980. *Note:* These were carried out at that time because of the known adverse effect this had on the MoC in the summer of 1980.

COMPARING YOUR RESULTS TO OTHER FARMS

Up to now the emphasis has been placed on the comparison of your results to your budget target. It is most important, however, that you should have a yardstick against which to measure your own budgets and performance. This is best done by taking part in some form of group costing scheme, but beware the dangers of 'cooking' the results so that they look good or of drawing the wrong conclusions due to the lack of authenticated data.

Try to avoid drawing too many conclusions from one month's figures. Ideally you also want to be able to see a whole year's results for the comparative farms so that the monthly results can be placed in perspective. *Keep cow numbers up to budget.* This may seem a simple thing to do but 'not having enough cows' is a main reason for lack of success on many dairy farms.

The number of cows cannot, of course, be looked at in isolation. 'Not enough cows' means that there is a shortage of cows relative to available feed supplies. If cow numbers are lower than budget because of lack of feed supplies that is another story.

To sum up we need to have at our finger tips the following information:

- Actual number of cows compared to planned number of cows.
- Actual yields of monthly calving groups compared to the previous year's results for similar groups, to targets and to results achieved by other farms.
- Actual levels of concentrate feeding compared to previous year, to our target and to results achieved by other farms.
- Service and conception records.

Methods of doing this are described in Chapter 12. Poor control now of this aspect of management will not show up in the financial results for another 9–12 months and by then it is much too late to do anything about it.

COW NUMBERS/FEED SUPPLIES

As was mentioned in the previous chapter it is most important to plan to have a reserve or carry-over of winter feed supplies. This will allow a late spring to be safely negotiated without recourse to the need to purchase expensive fodder supplies to eke out a shortage. Running short of feed in April is one of the *most certain* ways of ensuring that the budget targets are *not* reached. Most costing schemes commence in April and a poor result in this month tends to depress the results for the rest of the year.

Cow numbers/feed supplies is a basic and fundamental profit factor which must always be borne in mind when considering the merits of the alternative schools of thought in relation to grassland conservation, *i.e.*, whether to go for bulk or quality.

Whatever system you adopt it is important that come October you assess your winter feed supplies against budget and the targets you set earlier in the year and then make any necessary adjustments to your strategy. However, October may be too late. By end of July you should have a good idea of your likely feed stocks and this is the time to make decisions such as whether to grow stubble turnips or order additional brewers' grains. If you use sugar beet pulp you should have ordered enough last year to feed this year in September and October, *i.e.*, before this year's supply becomes available.

Unfortunately, assessing the quantity of feed supplies relative to the livestock requirements is not as simple as at first it might appear. Firstly the length of the winter period or date of turn out is not known within 30 days and this represents at least 15 per cent of the total requirement. Secondly the weight of silage in a clamp varies by up to 20 per cent whether one believes there is 0·65 or 0·8 tonne per cubic metre.

The above is particularly true on self-feed silage systems. A very useful tip in this case is to keep a simple record of the date each bay is completed. This will allow you to calculate the number of cow days per bay and will be valuable information in the year of shortage. There is no excuse, however, for 'running out of silage and/or hay' except in most exceptional circumstances. The number of days to 'turn out' should be calculated and the home-produced feed should then be rationed accordingly. On a self-feed silage system this may simply mean that the fence has to be moved 10 cm per day instead of 12 cm. This will reduce silage intake and consequently other feeds will need to be used. Whether this takes the form of purchased concentrates, brewers' grains, home-grown cereals, hay or other purchased feed will depend on individual circumstances.

Occasionally, despite the best laid plans, the winter may start with a definite shortage of fodder and this raises the question 'Should I sell some cows?'. But is this question the right one? If we have dairy replacements or beef cattle on the farm as well as cows the question should be broadened to 'Should I sell some cows or other cattle to eke out fodder supplies?'

To answer this question we need to be able to assess fairly accurately the quantities of feed that are required per animal. This will vary from farm to farm according to the feeding system. In this instance, therefore, it is proposed to try to answer the question by referring to the case-study described in the previous chapter.

The target on this farm is to produce 840 tonne (6 tonne per cow) for the dairy herd and 120 tonne for the yearling dairy heifers

Table 11.1. Example costs saved by selling heifers during fodder shortage

	£ Per heifer sold
(a) *October to April*	
Hay. 5 kg per day for 200 days = 1 tonne @ £50	50
Concentrates. 2 kg per day for 200 days = 0·4 tonne @ £110	44
Bedding straw 0·5 tonne @ £12	6
Vet. fees and livestock sundries	2
Share of costs of Hereford bull (£160 ÷ 40)	4
	106
add	
Sale value of bulling heifers (325 kg @ £1.00 per kg)	325
	431
add	
Interest charges saved. £431 @ 12 per cent × ½ year	26
Savings in labour and machinery costs, say	3
	460
(b) *May to October*	
Grazing variable costs—0·12 hectare @ £83	10
Interest charges on £460 for 6 months @ 12 per cent	28
	460
TOTAL	498

(3 tonne per head). This gives a total of 960 tonne and should yield a reserve of 140 tonne (1 tonne per cow) in a normal year.

In the year in question total supplies at the start of the winter are found to be only 770 tonne ($5\frac{1}{2}$ tonne per cow).

So, will it pay to buy *all* the feed for the dairy replacements as we need all the silage for the dairy cows?

To answer this question we need to compare the costs that will be saved plus the income received from selling the bulling heifers to the price we would need to pay in 12 months time for down-calving heifers (see Table 11.1).

The accumulated costs of the heifer by the time it calves is £498. The expected purchase price for an equivalent heifer in twelve months time is at least £500, and it is considered that there is a good chance it will be in excess of £550. It is concluded that the heifers should be reared despite the need to purchase all the food.

Note: If there had been adequate silage for heifers all the costs would still be the same as shown except the cost of food. This would be down by approximately £60, *i.e.*, all the hay plus $\frac{1}{2}$ kg concentrates per day. Not having reached the silage target represents a potential loss of profit of approximately £2,400 (120 tonne silage @ £20 per tonne).

Note once again that the finance costs change as a result of this decision although these are normally thought of as part of fixed costs. Turning back to the question of selling some cows, the partial budget would be along the lines shown below.

		£
Gross margin foregone by not keeping one cow		350–450*
Less cost saved	£	
Cost of replacing 6 tonne silage (as hay & concentrates)	120	
Bedding straw	10	
Interest charges £500 × 12 per cent	60	
Labour & machinery—difficult to assess as these will be much the same irrespective of cow numbers, say	20	
		210
Net loss due to selling cows		140–240

* According to efficiency of dairy herd

Note: The net loss resulting from not keeping cows is *much* greater than that from not keeping the dairy heifers. This is true on virtually all farms so in times of shortage whether it be of fodder or finance it is nearly always better to sell youngstock rather than cows. In times of shortage, however, your are likely to be one among many, so youngstock prices will also be depressed. Do your sums carefully if this is the case before you sell anything!

CASH FLOW BUDGETING AND CONTROL

There are two major objectives in cash-flow monitoring and budgeting. These are (i) to check that expenditure and revenue are being kept in control and proceeding to plan, and (ii) to determine the seasonality and size of the borrowed capital requirement.

The seasonality of the borrowed capital requirement is much less marked on dairy farms than it is on arable farms. The main reason for cash-flow monitoring on dairy farms therefore is to enable a tight control to be kept on revenue and expenditure, and to avoid taking on over-ambitious capital expenditure programmes.

(a) Preparing the Annual Cash Flow Estimate

To illustrate this we return to the case study described in the previous chapter.

The first step is to prepare the annual cash-flow estimates from the gross margin/fixed cost budgets detailed on Schedules 2, 3 & 5, and to set these out as shown in Schedule 7.

In most instances it is not necessary to break this down into more than four quarterly estimates. Monthly estimates may be necessary at critical times but the preparation of these can be very time-consuming. The cash-flow form illustrated in Schedule 7 is laid out so as to ease the separation of the trading and capital terms (see trading items totalled at lines 14 and 37). When preparing cash flows it is important to remember the difference between receipts and revenue on the one hand and payments and expenditure on the other. This is illustrated on Schedule 10.

For the sake of simplicity the budget receipts and payments shown on Schedule 7 are assumed to be exactly the same as revenue and expenditure so that these can be reconciled with the gross margin and fixed cost budget. In practice they will not be identical due to variations in the level of creditors and debtors between the beginning and end of the year; see again Schedule 10.

Note: The revenue and expenditure totals shown on Schedule 10 are on Schedule 11 in a more familiar form as part of a trading account. One general point to remember when preparing cash flows for dairy farms is that milk receipts, *i.e.*, the actual cash received, is one month in arrears of sales. The Milk Marketing Board has also recently adopted the principle of late payment for part of the milk to fund its capital projects so a small proportion of the cash for monthly milk sales is received three months in arrears, and a final payment is received twelve months in arrears.

The major assumptions made about timing of payments and

receipts when preparing the cash flow need to be noted; the assumptions made in respect of our case-study farm are shown on Schedule 8. What conclusion can we draw from the cash-flow estimates? (Please examine Schedule 9 and read the notes on Schedule 8 before continuing with this section.) Note in particular Item 14 on Schedule 8, that is the manager's wish to spend £20,000 as soon as possible on updating the parlour.

1. It would appear that it *may* be possible to fund the first half (£10,000) of the parlour improvements in the final quarter of the financial year (*i.e.*, in March 1982) without requiring an increase in the overdraft facilities compares to that required in March 1981, *but* funding this investment will mean that a significant reduction in the borrowed capital requirement cannot be made.

2. It would seem reasonable at this stage to expect a similar cash-flow pattern in the following year and the expenditure of a further £10,000 to complete the improvements would almost certainly lead to a need for an increase in the overdraft during the summer of 1982.

3. If a decision is taken to complete the improvements before 31 March 1982 a substantial increase in the overdraft facility will be required even if the results go according to plan.

4. The results may not be as good as expected so a decision is taken to *delay* the improvements until 1982 and to only undertake them then if the results are going reasonably well.

(b) Monitoring the Cash Flow
At this stage the reader is reminded that it has been assumed that we prepared the budgets and the cash-flow estimates shown in Schedule 7 in advance of the start of the financial year, *i.e.*, in February for a financial year starting 1 April. Hence the assumption in the cash flow that revenue and expenditure are identical to receipts and payments. We now move forward 4–5 months, *i.e.*, to July, and to the comparison of the actual results against budget for the quarter ending June 1981. These figures are shown on Schedule 9, columns A and B. Examination of these figures reveals several significant deviations from budget and these are summarised below.

	Line No.
(a) Concentrate payments are £6,000 more than budget	23
(b) The rent and rates are £3,500 more than budget	33
(c) The overdraft at the start of the year is £6,000 less than budget	47
(d) The overdraft at the end of June is £3,900 more than budget.	48

We now need to examine and find the reasons for these deviations. The rent and concentrate feed deviations are largely due to timing of payments. The rent was paid in the first week of April instead of the last week in March, and two months' feed bills were owed at the end of March instead of the usual one. If these two items had been paid as planned then the overdraft at the end of March would have been £33,500 instead of £24,000:

	£
Actual overdraft, end of March	24,000
add rent not paid	3,500
feed bill not paid	6,000
	33,500

We have therefore started the year with an effective overdraft requirement of £33,500 compared to our budget of £30,000. Hence the point made earlier that it is much easier to monitor cash flow if the budgets are prepared after the year in question has started.

The above items will confuse our comparison to budget throughout the rest of the year. The budget is therefore modified to take account of these factors in the completion of the *accumulative* cash-flow figures shown on Schedule 9 for the rest of the year; that is, the budget figures are adjusted to accommodate the errors simply due to the timing of payments at the *start* of the year. The total cash flow on rent for the whole year becomes £11,100 instead of £7,600, and feed is £63,600 instead of £57,600. But note that in practice these amendments do not need to wait until the end of the first quarter: these errors in timing will be apparent within one month of the start of the year and that is the time when the amendments should be made.

Having made our amendments we are now aware that, other things being equal, we shall have an overdraft requirement at the end of the year of £31,300 compared to our original estimate of £27,800 (see line 48, column G, schedule 9). In view of this we shall not be able to fund the parlour improvements unless we can beat our budget during the rest of the year.

Let us move forward now a further six months to January and the comparison of the assumed actual results for the 9 months ending 31 December to budget. These results are shown in column F, Schedule 9.

The trend in the actual results compared to budget is encouraging. Total trading receipts (line 14) are up by £7,770, more than offsetting the adverse variance in payments and giving an overdraft at the end of December which is £1,200 below budget. However, we are rather discouraged to note that most fixed cost items, see lines 30–35 and 38–39, are tending to run slightly above budget. Interest charges on the other hand are slighly below budget reflecting a favourable movement in interest rates.

The cows are milking well and the total receipts are 6 per cent up on budget. We have needed to cull rather fewer cows that expected which has allowed us to sell three more heifers than planned and has increased the net cattle receipts by £1,300 compared to budget. It is encouraging to note that the additional milk has been achieved without spending much more on feed than originally budgeted. (*Note:* We have checked the creditors to make sure that the actual feed payments are not low simply due to a delay in timing of payments.)

We are now in a position to update our budget to the end of the year, and our revised estimates of the whole year's results compared to budget are shown in column H on schedule 9. Total receipts are expected to be up by £10,000. This is offest by several adverse payments, particularly the need to spend more on fertilisers and lime but the overdraft at the end of the year is not expected to be more than £30,000 even if we spend £10,000 on the parlour improvements.

A year has now passed since we prepared our original budget and it is time to prepare our draft cash-flow budget for the coming year. This can be done quite readily given the information in Schedule 9 for the current year. Receipts and payments will tend to follow the same pattern as in the current year and this means that there will probably be a substantial cash surplus available to fund improvements during the third quarter, *i.e.*, October–December 1982. It can be arranged for this to coincide with the time when payment will come due on the parlour improvements. Things are going well and to target so we decide to go ahead and place an order for the improvements to be carried out in autumn 1982.

END OF YEAR MANAGEMENT ACCOUNTS

Time has moved on; it is May 1982 and it is only now that we have the information available to compare our results to the budget we prepared some fifteen months ago. At the end of March 1982 we carried out the annual stocktaking valuation. This should be entered on Schedule 1 alongside the actual valuation at the start of the year, *i.e.*, 31 March 1981, and also alongside our estimate of what we expected the 31 March 1982 valuation to be when we prepared our budget 15 months ago. The cash flow and revenue/expenditure results are shown on Schedule 10 and from this we have prepared a trading account summary as shown in Schedule 11. From this we can prepare gross margin/fixed cost results for the year ending 31 March 1982 and compare these to the original budgets we made at the start of the year, using Schedules 12 to 15. These results are also in a form that will also allow us to compare our results to 'standards' but we shall have to wait several months before these are available and by that time we shall be estimating our results for the year ending March 1983.

Chapter 12

IMPLEMENTING THE DAIRY HERD PLAN—MONITORING ENTERPRISE PERFORMANCE

MILK YIELD MONITORING

We start this chapter by repeating a statement made early in the previous chapter. 'The monthly MoC data tells how well you have done but if the results are not as good as expected it does not tell you the reasons why, and it also tends to be too late for remedial action.'

To overcome this problem a system of management control has to be developed which allows deviations from expected performance to be diagnosed promptly so that immediate remedial action can be taken. Many years ago when herds were small in size it was not uncommon to record milk yields daily but this practice became uneconomic with the increase in labour costs. Some farmers gave up milk recording altogether and most officially recorded herd owners changed to once-a-month recording as standard practice, instead of once per week, when this became permissible from a National Milk Records point of view.

The authors of this book consider that although once-a-month recording is acceptable for the National Milk Records, it is *not* sufficiently frequent for effective *Management Control*. They are also convinced that in most instances it is necessary to record feed use as well as the milk yields of individual cows if adequate control is to be exercised in the management of the herd. We go back to the first chapter in this context and repeat one of the questions asked there: what are we trying to do? Answer: We are are trying to feed the cow to produce a given yield. Are we achieving that yield? If the answer is yes, could we achieve that yield at lower cost? If the answer is no, what can we do about it?

If we have this kind of management by objective approach we

will set ourselves targets and then compare our results to those targets. The Brinkmanship Recording System described in this chapter developed from the authors' attempts to apply this management approach to the feeding of cows and in their attempts to put over this concept when teaching students.

Work done at the NIRD in the 1960s by Dr Broster and others led to the publication of standard lactation curves for dairy cows. These curves indicated that a cow should reach its peak yield in the 5th–6th week of lactation and that after reaching this peak the yield should not fall by more than $2\frac{1}{2}$ per cent per week. These lactation curves became the basis of control of the feeding of the cows and the emphasis in the initial years was on *how little concentrate feed* could be used consistent with yields not falling at greater than $2\frac{1}{2}$ per cent per week. As the years progressed it became evident that the then conventional system of feeding placed too much emphasis on concentrate economy because it was found that peak yields were being reached in the second and third week

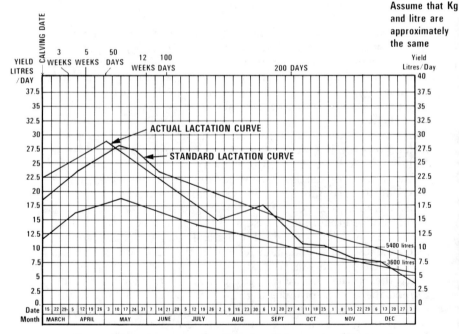

Fig. 5. Lactation curve for March-calving group of cows

of the lactation and these peaks were relatively low. As a result greater attention was paid to the feeding in early lactation with the objective of producing higher peak yields.

More concentrates and less bulk feeds were fed in early lactation than had previously been the case and this led to substantial increases in both the peak yield and total lactation yield. It also led to different shaped lactation curves for high-yield cows; these are much flatter than the standard curves.

BRINKMANSHIP RECORDING SYSTEM

In most dairy herds calving is spread over a wide period and at any point in time there are dry cows, cows in early lactation, cows in mid lactation, and cows in late lactation. The feed requirements of these cows at differing stages of their lactation vary widely as do their average yields. It is pointless, therefore, to study the average yield of the whole.

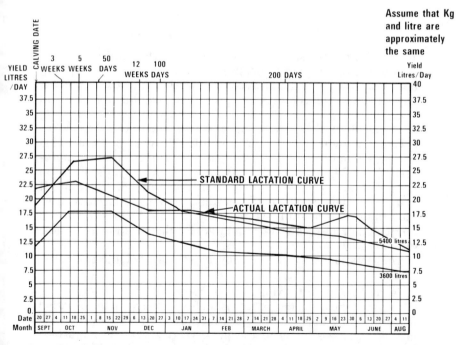

Fig. 6. Lactation curve for September-calving group of cows

The essence of the Brinkmanship recording system is that the milk yields of individual cows are recorded weekly on a group basis according to the month in which they calve. The average yield for the group is then plotted and related to feed input and to yield expectations. The plotting of milk yield in this way can reveal some startling results. Two examples of results that have been found on farms when results have been plotted *in retrospect* are shown on the lactation charts.

The first is for a March-calving group of cows. Milk yield in early lactation showed promise but yields fell away in June and July due to drought, to be revived briefly in September by a late flush of grass. In late lactation with the onset of a cold, wet, early winter, production again suffered as the cows were left outdoors too long.

The second is for a group of cows calving in September. The peak yield is low as the cows were out at grass in October. There is a small improvement when the cows are brought indoors but the main feature of the lactation chart is the substantial rise on turning out to grass.

LACTATION MONITORING

These illustrations show what can happen when no records are kept of performance or to be more accurate, records are kept but not used. The objective in recording weekly is to try to prevent unsatisfactory yields by taking appropriate action before it is too late.

The lactation charts shown have on them the so-called 'standard curves' for cows calving in these months. These are a useful guide but as mentioned earlier, these 'standard curves' are not appropriate for high-yielding herds as the standard curve is based on national results and these include herds that are 'mismanaged'. If you have a high-yielding herd you will be looking for a flatter curve than those shown and for a much higher peak yield, probably in the region of 40–45 litres with a target for the lactation as a whole of 8,000 to 9,000 litres.

The lactation curve can be conveniently divided into two parts.

1. Early Lactation: Calving to Peak Yield

This is the time when the pattern is set for the whole of the lactation and for a long time it has been generally accepted that the total lactation yield is equal to the peak yield × 200. The practical implication of this is that to attain a lactation yield of six thousand litres, monthly group must attain an average peak yield of thirty litres per head.

If the whole of the dairy herd budget is based on the expectation of a six thousand litre average, then the target is clearly defined for peak yield. The time taken to reach peak yield does vary and can be anything from six to twelve weeks depending upon factors such as level of feeding, cow condition and environment.

If the weekly recordings after calving show that a group average yield is not moving to the intended level, then there must be immediate discussion during the first two or three weeks to find out what is causing the failure to approach peak. Questions must be asked as to what can be done about it and possible remedies must be tried.

In most cases a failure to reach the target peak yield will be due to inadequate feeding and this aspect will be considered in detail later.

A most important task to be fulfilled is the service of the cow at the appropriate time to secure a good calving interval for the ensuing year. Service before three months in milk will secure a 365 calving index.

Decisions as to when a cow should be served are related to total herd management and range over many other factors such as yield levels, change in calving pattern, summer/winter milk policy etc., and will be discussed later.

Once a cow has been served successfully, then milk yield starts to decline and so we now move to a consideration of the second part of lactation.

2. Mid/Late Lactation: Peak to Dry

When peak has been attained and all cows in a monthly group put in calf, the monitoring of group average milk yields takes on a new significance.

To attain the budgeted peak yields it is highly likely that the group of cows will have been fed on a high energy/high protein ration. This is an expensive business and efforts during the remainder of the lactation should be directed toward reducing the cost of the ration whilst still attaining the desired level of milk yield and acceptable rate of decline.

If changes in ration are made the effect of this is monitored by checking the average group milk yield the following week. The objective is to try to relate cause and effect. Translating husbandry into economic terms, the two factors to balance are the possible loss of milk against the reduction in the cost of the ration.

As well as pointing the way to obtaining the milk yield more economically, regular milk recording also pinpoints changes which can occur in late lactation which are damaging to the level of milk production but which could be remedied.

On many farms, silage is made in two or three cuts. These can be analysed to give some indication of quality. The quality of the silage can only be effectively demonstrated by the performance of a group of cows when they move on to a different cut.

If the analysis suggests that the next silage cut to be eaten is poorer in quality then supplementary feed can be arranged and the change-over smoothly effected. Better silage would allow a reduction in the cost of the ration whilst still producing the same level of milk.

When milk yield is recorded on a regular basis for all the groups of cows within a herd, then it is possible to manage a herd effectively and ensure that: (i) the herd milk production is proceeding along the guidelines set in the budget, and (ii) that any problems are dealt with immediately and in some cases be avoided.

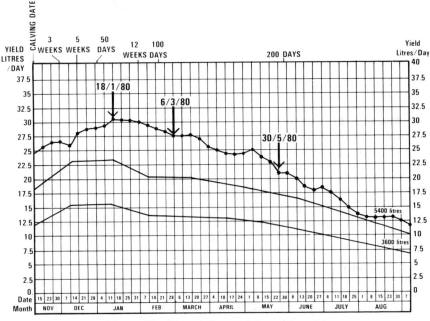

Fig. 7. The November-calving group of cows

EFFECTIVE HERD RECORDING—A CASE STUDY (see
Table 12.1 and Fig. 7)

In this example, information is provided to show the progress of
a group of twenty-one cows which calved during the month of
November 1979.

Table 12.1. Milk and feed records for 21 November-calving cows

Milk from bulk feed		$M^* - 5$		$M + 2.5$		$M + 20$	
Recording date		18.1.80		6.3.80		30.5.80	
Cow no.	Calving date	$Y^†$	$C^‡$	Y	C	Y	C
124	1.11.79	30	13	28	9	18	0
175	1.11.79	37	15	33	10	29	4
111	1.11.79	30	13	27	9	22	1
102	2.11.79	31	13	28	9	21	0
64	4.11.79	19	9	17	5	14	0
71	6.11.79	25	11	25	8	19	0
132	12.11.79	33	14	26	8	23	1
206	13.11.79	37	15	29	9	26	2
108	15.11.79	26	11	21	6	18	0
5	18.11.79	31	13	29	9	21	0
136	18.11.79	32	13	30	9	26	2
68	19.11.79	25	11	27	9	23	1
107	19.11.79	33	14	26	8	14	0
222	21.11.79	33	14	31	10	22	1
128	21.11.79	28	12	22	7	17	0
3	22.11.79	30	13	25	8	24	2
88	26.11.79	31	13	27	8	20	0
59	26.11.79	36	15	33	10	30	4
64	28.11.79	31	13	27	8	19	0
32	29.11.79	25	11	27	8	23	1
238	30.11.79	33	14	29	9	23	1
TOTAL		636	270	567	176	452	20
Average yield (litres)		30·2		27·0		21·5	
Ave. concs. fed (kg)			12·8		8·5		0·95
kg/litre			0·42		0·30		0·04

 * M = maintenance
† Y = yield per cow (litres)
‡ C = concentrates fed per cow (kg)

Time does not permit us to make a commentary of the milk/ feed information every week. However, three dates during the lactation have been chosen to illustrate how information can be collected, interpreted and turned into effective decisions.

The first step concerning herd management was to have analysed both the first and second cuts of silage; in this case the second cut was eaten first. The silage was made in an outside clamp in the Dorset wedge style. A self-feed system was in use.

From the analysis in Table 12.2 it can be seen that both first and second cut silage were of an inferior nature.

Table 12.2. Silage Analysis

| | 1st cut | | 2nd cut | |
	As received	Dry matter basis	As received	Dry matter basis
Dry matter g/kg	210·7		239·5	
Crude protein g/kg	27·2	129·2	35·7	149·0
pH	5·1		4·6	
MAD Fibre g/kg	86·5	410·4	89·8	375·0
Est. ME mg/kg		9·3		9·7
Est. DCP g/kg		81·9		98·1
D value		58·2		61

Given this analysis there could be little thought of obtaining milk from silage, and it would therefore have to be obtained from bought concentrate feed. This meant that the margin per litre would be low and high yield would be required to give a good margin per cow. The target peak yield was set at thirty litres and a policy decision was taken to try to avoid loss of bodyweight so as to secure a good conception to first service.

Table 12.1 shows the yield and feed input of these 21 November calving cows at three stages of lactation. These three have been chosen to show how changes were effected through the lactation in gradual stages, and the examples apply to recordings in January, March and June.

18th January 1980
At this point the peak yield of 30·2 litres was obtained and to achieve this a feed level of M − 5 litres had been applied, that is to say concentrates were fed for all milk produced plus five litres or alternatively silage was producing 'maintenance minus five

litres'. If the silage had been better then it might only have been necessary to feed at M − 2·5 litres.

In this herd, no physical grouping is possible but individual feeding is relatively easy as an out-of-parlour feeder is available. Feeding to yield gave all cows the opportunity to realise their genetic potential, whether high or low. A high energy, that is 0·35 kg per litre concentrate was fed and the average concentrate usage at 0·42 kg/litre would normally have been considered high at this time, but in view of the silage quality this was considered to be justified.

The intention had been to make good silage—whoever attempted to make bad silage? but unfortunately the attempt failed in this particular year. The adverse effect this had had on costs has been noted and is a spur to do better next time, if only the weather will help.

6th March 1980

By this time, with all the cows having been served, the concentrates were reduced to the extent that 2·5 litres were being taken from the silage. Yield had fallen to 27·0 litres.

The cows were now being fed on first-cut silage which on analysis proved poorer than second cut. However, the decline in milk was at an acceptable rate; the hope was that the cows would go to grass giving a sufficient yield to achieve 22·5 litres from grass.

The concentrate usage at 0·30 kg/litre was in line with the M + 2·5 level of feeding. Once again, this emphasises the importance of having a good-quality bulk feed for dairy cows; poor silage gives little milk. At first glance it would appear that this monitoring was only concerned with recording milk and feed. This is not so because the information builds up over the lactation, and in this case the fact emerged that the constraint in attaining higher margins was poor bulk feed. This means that investigation should be made into everything relating to the making of silage. System of silage making, including type of chop, degree of wilting, grass strains, level of fertiliser, must all come under the scrutiny and be discussed in the light of the fact that this year's silage was just not good enough.

Another feature which can be discussed is the variation in yield of cows within this particular monthly calving group.

The ability of cows to reach different peaks will have already been noted. By the end of the winter it is also evident which cows are able to hang on to their milk and which have the ability to produce milk from bulk feed.

Automatically cows will offer themselves for culling if they fail to comply with these two requirements. The matter of culling is not quite as straightforward as that: the availability of herd replacements, alteration to calving pattern and the cost of replacements all have to be taken into account.

By this time in the winter the main management questions should be: How much more milk should we attempt to take from silage, and how much will this calving group be giving when they go out to grass?

As it happened, another 2·5 litres were taken from silage during the last two weeks before turnout.

30th May 1980

The cows were turned out to grass on 20 April. The procedure was to allow them out for two hours, increasing to a full day during the first week. During this time, the cow's digestive tract becomes accustomed to the intake of grass and during this period little change can be made in the feeding of concentrates. During the second week after turnout, the cows are put out at night and the chase for milk from grass is on. We use the expression 'chase' with purpose because there is no time to be wasted. Good grass is usually only available for about eight weeks.

To obtain milk from grass there must be both quality and quantity available. The first is assured by early May but the second feature is very much dependent upon the weather.

Unfortunately, the turnout of 1980 coincided with a five-week drought which restricted the amount of grass available. The only way to make grass available would have been to sacrifice some of that set aside for conservation.

However, in this case, by the end of May, milk from grass was set at M + 20 litres; all cows giving twenty litres or less received no concentrates. It had been the policy on this herd to include calcined magnesite in the ration since before turnout to the end of May to counteract the possibility of grass staggers.

The attainment of twenty litres from grass is not unusual. In practice, however, the only way to get it is to cease feeding below that level. This is a positive objective policy and is the only way to make things happen.

As the summer proceeds so the quality of grass falls. It is a management decision as to whether there is a need for any supplementing of the grass.

Given a normal grass season, there should be adequate grass to sustain the normally declining milk yield until the cows are due to go dry at the end of August, prior to calving again in November.

With the detailed milk/feed information available each week, it is possible to work out the possible cost advantage, if any. If, for example, a group declines two and a half litres in one week at 12p per litre (= 30p) and this could be recovered by feed purchased for 15p, would this be worthwhile?

The answer would be governed by the businesslike instinct of the farmer. Only the higher-yield animals would need to be looked at in this way. The remainder of the group, giving lower yields, would milk through the summer with no supplement whatsoever.

FEED RECORDING

The ease or difficulty with which feed use can be measured varies greatly from farm to farm and depends upon several factors and in particular on the number of other enterprises on the farm and the ease with which bought-in concentrates can be allocated. Where home-grown cereals are grown and used on the farm, it is often difficult to keep an accurate account of the amount transferred in bulk from the arable to the dairy enterprise. Often there is no means of weighing cereals between harvesting and eventual feeding. Cereals that are sold are taken away in bulk consignments which are weighed accurately at the destination. The same discipline should be applied if at all possible to cereals used on the farm.

There has been a trend in recent years towards the transport of concentrates in bulk, rather in bags with obvious advantages in reduced handling charges both for the merchant and the farmer, but it must be realised that there is much less control over how much is fed when the feed supply is in bulk, unless some mechanical means is found for recording the use of the bulk feed accurately.

Surveys have shown in the past that parlour feeders have not been very accurate in dispensing feed. There has been much attention paid to this problem in recent years and now much more accurate dispensers are available for in and out of parlour feeding.

Another introduction has been the forage box with weigh cells, which again help to check on the accuracy of feed consumption.

From a management point of view it is important not only to check the accuracy of the dispensing of feed but to compare the level of feeding with what should have been fed. It is just as important that there should be no underfeeding as overfeeding. Any underfeeding, stemming from lack of control, can result in cow performance dropping simply because they are not being fed adequately in line with their requirements.

A Feed Record must be kept if there is to be any attempt at all to monitor the input of concentrate feed to the dairy herd.

A Feed Audit Sheet could be used in the simple form as shown in Table 12.3.

Table 12.3. Weekly feed audit sheet

	Concentrate type	
	High energy kg	Sugar beet nuts kg
Opening concentrates stock	5,000	2,000
Deliveries	5,000	2,000
Total available	10,000	4,000
Closing stock	3,000	3,500
Total concentrate usage in the week	7,000	500
Amount expected to use	6,800	480
Discrepancy AND WHY?	200	20

Purchased concentrates make up sixty-five per cent of Enterprise Costs; an attempt to maximise dairy margins must involve the close control of concentrate use.

MEASURING HUSBANDRY EFFICIENCY

The measures used to judge husbandry efficiency include:

- *Milk yield per cow in milk.* This is different from the usually accepted standard of milk yield per cow in the herd. The whole herd figure is useful for locating simple milk production levels.

  The yield per cow in milk reflects the day-to-day management and feeding of the herd. Any rise or fall in yield will then indicate whether the management changes were good or bad.
- *Concentrate use per litre milk produced per group of cows in milk.* The amount of concentrates used is a direct reflection of the quality of the bulk feeds. In any herd there will be a pattern through the year associated with its own particular seasonality of milk production.
- *Margin of milk over concentrates per day per cow, in each monthly calving group.* This is where the financial value of the milk produced and the concentrate cost of obtaining it can be combined to produce a margin figure. When calculated each

week this margin will give an objective answer to the question: What is the effect of change in the level of feeding?

Importance should be given to the relative change in margin from one week to the next.

Here is an example of a monthly calving group where in the late winter it was decided to take another 2·5 litres milk from the silage. The results of this action are shown in Table 12.4.

Table 12.4. Taking an extra 2.5 litres milk from silage

	Week 1		Week 2
Ave. milk yield/cow in group	22·3 litres		21·1 litres
Milk from bulk feed	M + 5 litres		M + 7·5 litres
Ave. concentrate use per cow in group	6·2 kg		5·7 kg
	£		£
Milk value 22·3 × 13 p	2·89	21·1 × 13 p	2.74
Concentrate cost 6·2 × 13·5 p	0·83	5·7 × 13·5 p	0·76
Ave. margin/cow in milk/day	2.06		1.98

In this case the reduction of concentrates resulted in a drop of yield rather greater than expected. The silage was not as good as we hoped.

The margin dropped by eight pence, this could perhaps have been rectified but the main fact to observe was that the yield dropped by 1·2 litres. This could reduce the potential of milk from grass in four weeks time when it was hoped that the cows would be turned out to grass.

The margin calculated directs attention to the present situation but thought also has to be given to the long-term implications of feed changes. If this margin is calculated on a regular basis for a group of cows calving in the different months of the year then after a couple of years it can begin to provide answers to the question: Which is the best month to calve on this farm?

These are many other factors to consider with block calving but the margin per monthly calving group would provide information that is currently not available from any source.

In the world of today, husbandry must be tempered with a highly developed economic sense so husbandry efficiency must be related to economics.

BREEDING AND VETERINARY RECORD

Mention was made previously of the importance of getting cows in-calf and the attainment of a good calving index.

The basis for success in this area of dairy cow management rests upon cow observation and identification. All cows must be capable of being easily recognised. This is now being widely appreciated and most cows now are either freeze branded or have a numbered collar.

The breeding cycle of the dairy cow follows a very regular pattern. It is necessary simply to follow the cycle for each individual cow in the herd and record events as they happen, as follows:

1. Observe that all the cow reproduction organs resume their normal shape and function at an early time after calving.

Retained afterbirths can provide problems which may need veterinary attention.

2. A normal, healthy cow will show signs of first heat at or about five weeks after calving. It is important to note this as it confirms that the cow is back on its breeding cycle.

3. The second heat period, twenty-one days after the first, offers the opportunity to serve the cow. Whether this option is taken up is a matter for discussion, taking into account other factors such as level of milk yield, body condition and the future intended calving pattern.

4. If the cow was served at eight weeks, then it is important to observe whether the animal returns to service after another three weeks. If not, it can be assumed that the animal is in calf.

It is possible to have a milk sample tested by the MMB twenty-eight days after service. Alternatively, the cow can be pregnancy tested by a veterinary surgeon. This test is usually performed some eight to ten weeks after service.

Once the pregnancy has been confirmed the calving data can be established and the appropriate drying-off date recorded.

When the herd breeding cycle has been so organised it is equally necessary to set up a system to monitor the health of individual cows in the herd.

There are varying levels of sophistication available, but at the very simplest level a card should be kept which will detail all illnesses and treatments during the cow's productive life.

Items to be recorded include:
(i) incidence of failure to breed, retained afterbirth, difficult calvings, etc;
(ii) incidence of mastitis, location, treatment;
(iii) occurrence of milk fever and treatment.
As this type of information builds up on each cow record, it can be used to help decide which cows to cull or retain.

Breeding boards are now in use on many farms and provide a very useful aid. The one thing which they fail to do is to maintain a record of documented facts. It is desirable to have a card index for all cows so that breeding failures, veterinary treatments and all herd health details can be registered throughout an animal's productive life.

Chapter 13

CAPITAL REQUIREMENTS IN DAIRY FARMING

CLASSIFICATION OF FARMING CAPITAL

The word 'Capital' has a precise meaning to economists and is regarded as a factor of production separate from land. It is also kept distinct from money which is simply a means of exchange and a symbol of capital. A dairy farmer's capital is usually expressed in £p but the actual farming capital is the land, buildings, livestock and other assets in his possession. In this chapter the word 'Capital' is used in the popular sense, *i.e.*, as a measurement of the £p required to buy or rent a farm and to farm it effectively. Later in this chapter, however, the reader will be reminded of the need to think of capital in more than £p terms.

The capital invested in agriculture is often classified according to the systems of land tenure, *i.e.*, into 'landlord's capital' and 'tenant's capital'. This served a very useful purpose when the vast majority of land was farmed on a tenanted basis. Today, however, this is not the case because over 70 per cent of land is owner-occupied.

It is also usual to divide capital into two classes according to whether it is 'fixed' or 'working (circulating)' capital. Fixed capital is that which does not wear out quickly and is replaced infrequently, such as buildings and fixed equipment. Working or circulating capital is that which is used or consumed in a single process and is replaced once per annum or more frequently, *e.g.*, money required to pay wages and purchase seeds and fertilisers. This simple division, however, is not convenient in farming as it is difficult to know where to place items such as dairy cows and other livestock.

We need to bear the above classifications in mind but in farming it is better to adopt a classification based on the nature of the goods and services and whether one would regard the capital investment as long-term, medium-term or short-term, as shown on page 167.

1. *Long-Term Capital Investments*
- Land
- Buildings
- Other fixed equipment

2. *Medium Term Capital Investments*
- Breeding livestock
- Machinery and equipment

3. *Short Term Capital Investments*
- Non-breeding livestock
- Seasonal working capital, including seeds and fertilisers, cost of cultivations and crop produce.

Table 13.1. Capital requirement—80 hectare farm

CAPITAL INVESTED

1. *Short term*	Total £	per hectare £	per man £
Debtors	14,500	180	
Saleable crop produce	Nil	—	
Purchased store	10,200	127	
Cultivations and tenant right	10,000	125	
Other crop produce	2,800	35	
Non-breeding livestock	20,400	255	
	57,900	722	19,300
2. *Medium Term*			
Machinery and equipment	28,000	350	
Breeding livestock	56,600	708	
	84,600	1,058	28,200
3. *Long Term*			
Buildings and fixed equipment	Nil		
Land	Nil		
TOTAL INVESTMENT	142,500	1,780	47,500
CAPITAL BORROWED			
Trade creditors	14,500	180	
Bank overdraft	30,000	375	
Other borrowings	Nil		
TOTAL BORROWINGS	44,500	555	14,833
NET CAPITAL INVESTED	98,000	1,225	32,667

CAPITAL REQUIRED TO OPERATE A DAIRY FARM

A good indication of the capital required to operate a farm can be obtained from an examination of the data shown in Schedule 1 (Appendix 3) for the case-study farm described in the previous chapters. This is summarised in Table 13.1. The farm is 80 hectares in size and employs three men.

Before commenting on the figures we should be aware of the need for a rather more precise definition of terms. What do we mean by 'Capital Requirement'? Do we mean the 'total investment' in the farm or the 'net capital investment'?

The total capital investment required to operate the farm is £1,780 per hectare (£1,600 excluding trade debtors). The net capital required to be provided by the owner of the business is reduced to £1,225 per hectare due to his ability to borrow the difference.

BREAKEVEN BORROWED CAPITAL REQUIREMENT

The budget profit from the above farm is £16,900 or £211 per hectare after paying £5,100 interest charges, *i.e.* 17 per cent on £30,000 (see Schedule 2). The total sum that this farmer could make available to fund finance charges is £14,000 assuming that his personal drawings and tax payments on the previous year's profits total £8,000 (see Table 13.2).

Table 13.2. Funding finance charges

	£	£ per hectare
Budget profit margin	16,900	211
Budget finance charges (£30,000 @ 17 per cent)	5,100	64
Budget profit before finance charges	22,000	275
less personal drawings requirment, say	8,000	100
Available to finance borrowings	14,000	175

£14,000 will fund a total borrowed capital of £100,000 at 14 per cent assuming no principal repayments. However, one needs to allow for repayement of principal as well as interest charges and it would be much more prudent to put the maximum capital sum that could be serviced by £14,000 at £70,000 or £875 per hectare.

This gives a break-even budget as shown in Table 13.3.

Table 13.3. Break-even budget

	£	per hectare
Budget profit before finance charges	22,000	275
less personal drawings and tax	8,000	100
	14,000	175
£70,000 @ 20 per cent including principal and interest	14,000	175
	NIL	—

The performance levels assumed on this farm are well above average. Not many farms are capable of servicing a rent of £87.5 per hectare plus £175 per hectare finance and principal repayments. Nonetheless it is an indication of what can be achieved given a high level of management efficiency. It should be noted, however, that the farmer in this particular case study is hoping to use his surplus profits to fund improvements to his milking parlour. These improvements would not be feasible if he had to use these funds to service higher borrowed capital commitments.

A third point to note is that the total borrowings by the case-study farm of £44,500 are £13,400 less than the short-term capital invested in the business. If there is an unexpected fall in profits these short-term assets can be reduced in size and a corresponding reduction can be made in the borrowed capital requirement, *e.g.*, by delaying the timing of purchase and payment for fertiliser or by selling non-breeding livestock. An increase of say £40,000 in the total borrowings would raise the total well above that of the short-term assets and would make the business 'over-borrowed'. Great care would need to be taken to see that this capital was borrowed on a medium-term basis to avoid being placed in the position of having to sell longer-term assets such as dairy cows, the result of which would be financially disastrous.

CAPITAL REQUIRED TO OPERATE AND OWN A DAIRY FARM

The capital required to purchase a dairy farm at the present time is in the region of £4,000–£5,000 per hectare depending of the location and quality of the farm. If we take a figure mid-way between these two—£4,500—it is interesting to note this is approximately two and half times the investment required to operate and run the farm. If the farm is effectively managed the £1,750 per hectare invested to run the farm can be expected to show a profit, after rent, in the region of £275 per hectare, representing a return of 15–16 per cent on the capital invested.

The owner-occupier does not pay a rent, so after allowing for ownership expenses he could expect a total profit in the region of £227 per hectare instead of £275, on a total investment of £6,275 per hectare (£4,500 land ownership plus £1,775 operating capital), giving an overall return of approximately 5 to 6 per cent. At first sight this return appears derisory but is must be remembered that investment in land is a very effective hedge against inflation. It also needs to be borne in mind that the vast majority of farmers have owned their land for a very considerable time and its purchase or inheritance value would be much lower than its present-day value. A large proportion of the land has also been purchased at sitting tenant valuations which are only 50–60 per cent vacant possession values.

The question is often asked, What is land worth? Not in the simple sense of £p but in terms of what it is capable of producing. This is difficult to answer but one way of attempting to do this is to compare the price of land to other farm inputs. As dairy farmers the obvious comparison is between the price of land and the value of a dairy cow because one of the fundamental decisions in farming is the choice between investment in more dairy cows or more land.

Reliable figures for the price of land and the value of dairy cows are difficult to obtain but it is considered that Table 13.4 gives a fair indication of how these have changed over the last ten years.

In 1971 and 1972 10 hectares of land could be purchased with the sum realised from the sale of 43–44 newly calved dairy heifers. Land prices rose rapidly in 1973 and 1974 and the number required more than doubled to 92. Land prices fell in 1975–6 and at this time the number required was reduced to 54.

Land prices rose rapidly in 1979 and again the number of heifers required to purchase land was at a high level. At the present time (1981) land prices are no more than they were two years ago. Over this same period there has been an increase in the value of newly calved dairy heifers so only 77 are required to purchase 10 hectares of land compared to 89 two years ago.

At the start of this section on land prices it was stressed that reliable figures for the price of land and the value of dairy heifers are difficult to obtain. They both vary considerably according to quality and prices for whole years conceal wide seasonal variations. The point in including Table 13.4 is to demonstrate that the ratio between land prices and dairy cow values does vary. The astute dairy farmer is aware of this fact and tries to plan the expansion of his business in such a way that land purchase takes place when the dairy cow to land price ratio is favourable. He also has a rule-of-thumb way of measuring the value of land. If it takes more than 7 cows to purchase one hectare of land it is dear, if it takes less than 6 it is cheap, the average being 6 to 7.

The table also shows the value of dairy heifers relative to the minimum agricultural wage. The variation in this ratio is not as high as is the case with land. Dairy heifer values were high in 1972 and 1973 in relation to wages reflecting the severe shortage and consequently high prices for cattle at that time.

Table 13.4. Ratios between land values and dairy cow prices 1971–81

January–March year	Land value £ per hectare	Newly-calved Friesian dairy heifer value £ per head	No. heifers required to purchase 10 hectares	Weekly* wage £	No. weeks' wages to purchase one heifer
1971	500	115	43	14.80	7·8
1972	750	170	44	16.20	10·5
1973	1,500	210	71	19.50	10·8
1974	1,750	190	92	21.80	8·7
1975	1,400	220	64	28.50	7·7
1976	1,500	280	54	36.50	7·7
1977	2,000	310	65	39.00	7·9
1978	2,750	405	68	43.00	9·4
1979	4,000	450	89	48.50	9·3
1980	4,000	480	83	58.00	8·3
1981	4,000	520	77	64.00	8·1

* Minimum Agricultural Wage as determined by Agricultural Wages Board.

CLASSIFICATION AND SOURCES OF BORROWED CAPITAL

Credit requirements are classified in the same way as the need for capital investments, *i.e.*, long term, medium term and short term.

The main sources of credit for farming are summarised below:

1. *Long-Term Loans* for land purchase and major capital improvements to buildings and fixed equipment
   - Agricultural Mortgage Corp. Ltd.
   - Joint stock banks.
   - Farmer's relations.

2. *Medium-Term Loans* for machinery and equipment, minor capital improvements, and major expansion of livestock numbers.

- Joint stock banks.
- Hire purchase companies.
- Machinery and equipment leasing companies.

3. *Short-Term Credit Facilities* to cover seasonal variation in capital requirements for seeds, fertilisers and livestock.

- Overdrafts from joint stock banks.
- Merchants' credit.
- Special finance schemes financed by banks.

The Agricultural Mortgage Corporation only grants loans on the 'mortgage security' of property so loans from this organisation are only available to owner-occupiers. Loans are granted for periods up to forty years but most loans are taken out for periods not exceeding twenty-five years. A case for borrowing for a period longer than 15–25 years is difficult to make unless one is able to borrow at the start at a *low fixed* rate of interest and considers that interest rates are not likely to rise. A farmer who borrowed at a fixed interest rate of 7 per cent some twenty years ago is making considerable savings compared to the interest rate charged at the present time.

Borrowing for periods longer than 20–25 years is not recommended under normal circumstances as the effect on the annual repayment is quite small, as shown in Table 13.5.

Table 13.5. Annual repayment instalments for a loan of £100,000 at a fixed rate of interest of 15½ per cent

Term of years	Annual repayment £	Incremental saving(*) £
10	20,000	
15	17,350	2,650
20	16,330	1,020
25	15,890	440
30	15,680	210
35	15,590	90
40	15,540	50

* *i.e.*, As a result of extending repayment period by 5 years

To illustrate the above more clearly, let us assume that the £100,000 loan is for the purchase of additional land and that this will be the only long-term borrowing on a farm totalling 160 hectares.

By moving from ten to fifteen years we reduce the annual repayment by £2,650, equivalent to £17 per hectare, and this represents a significant saving. Moving from twenty to twenty-five years reduces the annual payment by £440, or £3 per hectare. This is a somewhat insignificant saving in relation to the total, as shown in Table 13.6.

Table 13.6. The effect of changing the repayment period

Term of years	Annual repayment £	'Rent equivalent per hectare' for a 160 hectare farm £
10	20,000	125
15	17,350	108
20	16,330	102
25	15,890	99
30	15,680	98
40	15,540	97

The term 'Rent Equivalent' has been used above to cover the loan service charge per annum. This is a very useful rule of thumb to use when considering the viability of a proposed land purchase transaction. If the annual repayment is more than the 'going rent' the viability of the proposition will need to be examined very carefully.

Loans can be taken up from the Agricultural Mortgage Corporation on a *fixed* or *variable* basis. In the case of the former the interest charged remains fixed throughout the term of the loan but the variable rate changes in line with changes in other interest rates.

The normal method of repaying a mortgage is by the 'annuity' method, *i.e.*, by *equal* half-yearly or annual instalments of principal and interest. In the first year virtually all the annual instalment is interest, but by the last year the position is reversed and the annual instalment is mainly principal.

This system gives a much lower first year's payment than does a loan repayble by equal instalments of principal and for this reason is usually the preferred method of repaying a long-term loan. This is illustrated in Table 13.7 for a long-term loan of £100,000 repayable over twenty years at 15·5 per cent rate of interest.

More interest is payable of course over the whole period of the loan on the annuity method but the main criterion to adopt when taking up long-term loans is to ensure that the loan can be serviced in the initial years. If these early years can be navigated, then the later years tend to take care of themselves.

Table 13.7. Loan repayment methods

	Annuity Method			Equal Instalments of Principal		
	Interest £	Principal £	Total £	Interest £	Principal £	Total £
Year One	15,500	830	16,330	15,500	5,000	20,000
Year Two	15,371	959	16,330	14,725	5,000	19,725
Year Three	15,223	1,107	16,330	13,950	5,000	18,950

An alternative way of repaying the principal is by taking out a 'With Profits Endowment Policy' which at is maturity will provide a capital sum equal to the value of the loan. This method has tax advantages but it suffers from the disadvantage of higher total payments in the initial years of the loan. It does provide life cover and is promoted actively by insurance brokers but a decision to fund borrowing repayments for farm purchase by this method should not be taken without careful consideration of the alternatives.

Considerable time has been taken up in considering the role of the AMC Ltd in long-term finance but the joint-stock banks are by far the most important source of funds for farmers. All the main banks promote their services for farmers actively and several have teams of agricultural specialists.

By tradition the banks are only interested in short- and medium-term lending but all the banks now offer long-term loans for farm purchase, competing actively with AMC Ltd in this market. Banks are very keen to lend to farmers, particularly owner-occupiers, because their security is excellent. There is little or no risk of the banks losing the money they lend to owner-occupied farms because the sums that can be serviced and therefore lent to such farms represent a relatively low proportion of their total assets. Earlier in this chapter, for example, it was shown that an owner-occupier is likely to have total assets in the region of £6,250 per hectare but can only service the finance charges on borrowings totalling £1,000–£1,500 per hectare which represents less than 25 per cent of his total assets.

Owner-occupiers, therefore, have no security problems when it comes to borrowing money. In many respects this is a disadvantage because it makes it relatively easy to borrow too much.

A cardinal rule when borrowing money is not to borrow for long-term needs using short-term funds. It is also essential before borrowing to prepare budgets and cash-flow estimates to ensure that the service charges can be met.

The most frequent borrowing mistake is over-capitalisation in the long- or medium-term at the expense of short-term requirements. It is not all that unusual to find a dairy farmer who has invested borrowed capital in new buildings and equipment and who is unable to find the funds to purchase the cows to utilise the new facilities. Lack of cows means lack of cash flow to service the borrowings and the final result of this fundamental management error is the need to sell the farm.

One of the biggest problems in the management of an owner-occupied farm is to reconcile the needs of the farm for capital expenditure with the actual capital resources available. On many farms the need for capital expenditure can be seen quite clearly but the means to repay this expenditure once it has been incurred are difficult to find.

Well-managed owner-occupied farms should have relatively little need to seek funds other than from AMC Ltd and the joint-stock banks unless they are 'over-borrowed'. This, however, is not the same as saying they do not make use of funds from other sources.

Tenant farms, however, have less security and unless they are well established need to turn to other sources for their borrowed capital, particularly for machinery.

Borrowing from other sources is nearly always more expensive than borrowing from the banks. It is generally accepted that an overdraft is the cheapest method of borrowing and if possible the bank should be the source of both short- and medium-term finance.

When you borrow capital there is always the danger of finding yourself short of funds at a critical time and with the consequent need to sell cattle or possibly grain at the wrong time. Selling at the wrong time—*i.e.*, when prices are low—has a disastrous effect on profits and financial plans should be prepared so as to mitigate the dangers of this happening. Cash-flow forecasts and other financial forecasts should include a contingency for unforeseen factors.

Chapter 14

THE NEW ENTRANT—GETTING A START IN DAIRY FARMING

LEARNING THE JOB

To succeed in dairy farming you have to be good at the job of milking cows and caring for cows. To acquire the necessary skills there is no substitute for practical experience. This practical ability to do the job on the farm has to be present if you want to work your way up the farming ladder.

A lot can be learned from watching how other people approach and do the job and a young person needs to avoid simply getting the same experience ten times over. This can only be achieved by working on more than one farm and this need to work on another farm is particularly vital in the case of a farmer's son.

This ability to do the job on the farm has then to be coupled with the ability to run the dairy herd and the farm as a business as has been described in this book.

CAPITAL REQUIREMENT AND SIZE OF FARM REQUIRED TO START FARMING

It needs to be stressed that the amount of capital and consequently the size of farm needed to create a viable one-man unit depends on the managerial ability and general capability of the individual. This is illustrated in Table 14.1.

Given above-average performance it is feasible to produce a margin from twenty hectares that is more than that produced at average performance from twenty-four hectares.

The data given in Table 14.1. indicates that it is possible to produce a viable full-time farm business in milk production from a farm in the region of twenty to twenty-four hectares. On a farm below this size it is very difficult to produce an adequate living without a non-land using enterprise.

An indication of the amount of capital that would be required for a farm of this size and how this would be invested is shown in Table 14.2.

176

Table 14.1. Expected profit margin on a 20–24 hectare dairy farm

| | Average ability | | | Above-average ability | | |
| | | Gross margins | | | Gross margins | |
	No.	per head	Total	No.	per head	Total
Dairy cows	48	350	16,800	40	420	16,800
Young stock	30	100	3,000	25	120	3,000
	hectares	*per hectare*		*hectares*	*per hectare*	
	24	825	19,800	20	990	19,800
Fixed Costs						
Power and machinery		275	6,600		275	5,500
Rent and rates		175	4,200		175	3,500
Sundry overheads		100	2,400		100	2,000
		550	13,200		550	11,000
PROFIT MARGIN BEFORE FINANCE AND LABOUR CHARGES		275	6,600		440	8,800

Table 14.2. Capital required for a 20 hectare dairy farm

	£	£ per hectare
ASSETS		
Machinery and equipment	10,000	500
Tenant's fixtures	2,000	100
Dairy cows	20,000	1,000
Dairy young stock	5,000	250
Hay, straw and silage	1,000	50
Cultivations and tentant right	500	25
Feed, seeds and fertilisers	2,000	100
	40,500	2,025
Debtors (one month's milk)	2,500	125
Total assets	43,000	2,150
LIABILITIES		
Creditors	3,000	150
Bank overdraft	4,000	200
	7,000	350
NET WORTH	36,000	1,500

The estimated figures are of course only a guide, the actual amount depending as already stated on the capability of the individual.

One factor that can vary enormously from farm to farm is the amount required as ingoing, particularly tenant's fixtures and equipment.

The estimates in Table 14.2 only take into account the farmer or tenant's capital requirements.

The capital required to purchase a farm at the present time is £4,000 to £5,000 per hectare, that is some two to three times the amount required for farm capital.

The would-be entrant to dairy farming wishing to purchase a viable one-man dairy farm will therefore need to be able to fund total assets in the region of £120,000 to £175,000.

	20 hectares	25 hectares
Purchase price: 20–25 hectares @ £4,000–£5,000	£80,000	£125,000
Farming capital, say £2,000 per hectare	£40,000	£50,000
Total capital requirement	£120,000	£175,000
Funded by borrowed capital, say £800 per hectare	£16,000	£20,000
NET CAPITAL FUNDED BY OWNER	£104,000	£155,000

ACQUIRING THE NECESSARY CAPITAL

1. *Inheritance.* Most people starting up in farming will only be able to contemplate the possibility because they inherited sufficient capital to give them the opportunity. If you don't have rich relatives or parents then you have to think up some other way. This probably means savings but you may be one of the lucky ones and achieve it by matrimony!

2. *Savings.* Although at first sight it may not seem feasible, people still manage to start farming when the basis of their capital provision is savings, not inheritance.

The ability to save is a prerequisite for a young person wishing to start and to be successful in farming as in his first few years he or she will almost certainly have to take considerably less out of the farm in terms of cash than the value of work they put in!

Inflation reduces the value of cash savings so it is important to know how to invest your savings as well as being able to make them. Buying a house is probably one of the best ways but this is not possible for most herdsmen or herds managers because they will probably be required to live on the farm. They need therefore to think of alternative investments. In this connection, however, it is interesting to note that the capital sum required to farm or as tenant is of roughly the same magnitude as the amount required to purchase a house.

3. *Part-time Farming.* This is one possible way of putting savings to good use providing that land and/or buildings can be obtained to put this into operation. The danger with part-time farming is the effect it may have on your job. Employers become rather anxious when they see their herdsmen dashing off to attend to their own enterprises.

Some form of part-time venture is very desirable, however, before anyone starts on their own as it gives a valuable insight into the problems of business and soon brings home the nature of the *risks* inherent in any business operation. Losses are made as well as profits!

4. *Share Farming.* This is not practised to any great extent in this country but is widespread in New Zealand.

It is considered, however, that some form of sharing farming could provide a very satisfactory way in which a young person could accumulate capital during his employment and do so in such a way as to benefit his employer. A way in which this can be arranged to the satisfaction of both employer and employee is described below:

1. The employee purchases up to 10–15 per cent of the cows in the herd at an agreed market price, preferably by introducing his own down-calving heifers.
2. Calves born to the employee's cows are credited to his/her account less marketing cuts and costs of getting the cows in-calf.
3. The employee receives 15 per cent of the Margin Over Concentrate Feed Cost per Cow per month multiplied by the number of cows he has in the herd.
4. The employee's likely share of profits relative to capital input is shown below:

	£
Calf output (net costs)	40– 50
	40– 30
Less herd maintenance cost	0– 20
15 per cent MoC £480–£560	72– 84
Total return	72–104*
Capital invested	500
Return	14–21 per cent

* Excluding appreciation in dairy cow unit value

5. The employer benefits through the greater attention given to all the cows in the herd as a result of his employee's increased motivation. The employer has a lower capital investment and lower bank borrowings. He also has the satisfaction of knowing that his employee is less likely to leave, but for the scheme to work effectively the employer must retain ownership of a large proportion of the herd.
6. Heifer calves that the employee wishes to retain can be contract reared by his employer.

ACQUIRING THE FARM

The lack of farms to let is the biggest stumbling block to the would-be farmer where parents are not farmers, but the enlightened development of schemes such as that outlined above may provide an alternative entry.

At first sight there may appear to be no problem for the farmer's son or daughter but in this instance there can be problems in retaining the farm, particularly in the case of owner-occupiers. The total capital invested in an owner-occupied farm is in the region of £5,000–£6,000 per hectare giving a total of £400,000–£480,000 for a typical eighty-hectare family farm. If there are two or more sons and daughters and all want to farm/or one wants to farm and one or more wants to take their eventual share out of the business this can lead to completely unviable units.

Much is written about the problems of a young person getting a start in farming or being able to continue in farming and the question of tenancy legislation is very topical at the present time. The present arrangements virtually guarantee succession for tenant farmers' sons. This is very satisfactory from their point of view but this virtual guarantee for them does mean that there are fewer opportunities for others.

When considering prospects for new entrants to dairy farming one has also to take into account the decrease in the number of dairy farms. In 1980 there were only 43,358 registered milk producers in England and Wales compared to 123,137 in 1960. This concentration of milk production into the hands of fewer producers has led to a substantial increase in herd size. The average herd size has gone up from 21 to 63 over the same period and in 1980 some 55 per cent of all cows were in herds of more than 70 cows. As mentioned in Chapter 1 this increase in herd size has led to many more openings for herdsmen and herd managers. In this sense therefore the prospects for a new entrant to dairy farming—that is of obtaining a well-paid and satisfying job in dairy farming—are much better now than they were twenty years ago. This has to be weighed against the increased difficulty the new entrant has in trying to start up in business on his own.

FUTURE OUTLOOK

This book has been written at a time when dairy farming and the country as a whole is in economic difficulties and the financial data quoted reflects this situation. It is essential, however, to remember that dairy farming like other aspects of farming is subject to wide fluctuations in profitability and too much note should not be taken of the bad times unless one feels there has been a *basic* shift in the fundamental economics of farming. One also has to remember that relatively small changes in the supply of agricultural products can quickly lead to a surplus becoming a shortage due to the basic inelasticity of demand for food.

Attention has been drawn earlier in the book to the dependence milk production profitability has on the profitability of beef production. At the present time (1981) beef prices are firm despite the general economic depression reflecting a relative shortage in supply of beef. This in turn is dependent on many factors but one of the most fundamental is the number of cattle kept on farms in this country.

Cattle numbers at the present time (1981) are lower than in the mid and late 1970s and are only at approximately the same level as they were prior to the beef boom years of 1972–3. This would suggest that there is a reasonable chance that beef will be in short supply when the national economy as a whole begins to improve which one presumes it will do sooner or later, possibly in 1982–3. This could lead to a substantial rise in beef prices and is one reason why the authors are optimistic about the prospects for dairy farmers in the next few years.

Despite this optimism it is thought that economic pressures will continue to force out the less efficient producer and that the number of milk producers will continue to fall. At the same time it is expected that yields per cow will continue to increase and the rate of increase in the next few years may well be greater than in the past due to the action being taken by producers to overcome their present economic difficulties.

Returning to the prospect for the new entrant, a slight easing of the problems involved in getting a tenancy may emerge as a result of the proposals being put to the government by the National

Farmers Union and the Country Landowners Association but the competition for these tenancies will still be severe. Against this can be set the prospect of good career jobs in dairy farming and it is up to the new entrant to decide at which goal to aim. Only he or she can answer the questions:

> What am I trying to do?
> What is stopping me?
> What can I do about it?

APPENDICES

1. Trading Account and Valuation Details
2. Terms and Definitions used in Farm
   Business Management
3. Case Study Farm Schedules

Appendix 1

Trading Account—Expenditure

		No. bought (if possible)	A	£ Total		
LIVESTOCK PURCHASES	Dairy cows and calved heifers		53			
	Bulls for dairy herd		54			
	Dairy heifers (including calves)		55			
	Beef cows		56			
	Bulls for beef herd		57			
	Other beef cattle (including calves)		58			
	Ewes for breeding and rams		59			
	Store lambs and other sheep		62			
	Sows and boars		63			
	Other pigs (incl. in-pig gilts)		65			
	Poultry		66			Sub-Total
	Other livestock (specify)		67		68	
FEED	Bought concentrates		69			
	Bought hay and feeding straw		70			
	Other feed		71			Sub-Total
	Keep rented		72		73	
OTHER VARIABLE COSTS	Fertilisers (£) Lime (£)		74			
	Purchased seeds and plants		75			
	Crop sprays		76			
	Crop sundries (incl. P.M.B. Levy)		77			
	Veterinary and medicines		78			
	Contract rearing: cattle		79			
	Contract rearing: sheep		80			
	Livestock sundries (incl. purchased bedding)		81			Sub-Total
	Other		82		83	
WAGES AND SALARIES	Regular labour (incl. national insurance)		84			
	Casual labour		85			
	Wife (if paid)		86			
	Paid management		87			Sub-Total
	Secretarial		88		89	
POWER AND MACHINERY	Machinery and vehicle repairs		90			
	Contract and transport		91			
	Leasing charges		92			
	Fuel and oil		93			
	Electricity, coal, gas etc.		94			
	Vehicle tax and insurance		96			Sub-Total
	Depreciation of cars, vans, lorries, tractors and other machinery		97		99	
PROPERTY CHARGES	Rent		100			
	Rates		101			
	Depreciation of buildings, improvements and fixed equipment	Tenant's items only See Note 10	102			Sub-Total
	Repairs to buildings, fences etc.		104		105	
SUNDRIES	Water		106			
	General insurance		107			
	Office and telephone		108			
	Professional fees (accountant, etc.)		109			Sub-Total
	Others (incl. subscriptions etc.)		110		112	
FINANCING CHARGES	Bank charges and interest		113			
	Mortgage and other loan interest		114			Sub-Total
	Hire purchase interest		115		116	
	Total Expenditure (box 68 + 73 + 83 + 89 + 99 + 105 + 112 + 116)				117	
	Opening Valuation (box 52)				118	
	Total (box 117 + 118)				119	
	PROFIT (box 185 less box 119)				120	
SUPPLEMENTARY INFORMATION	Value of unpaid manual labour (incl. farmer)		121			
	Estimated rental value of owner-occupied land		122			
	Area of owner-occupied land (hectares)		123			

Trading Account—Revenue

		No. bought (if possible)	£ Total		
LIVESTOCK SALES (include associated premiums and subsidies)	Dairy cows		124		
	Bulls from dairy herd		125		
	Calves from dairy herd (0 to 3 months)		126		
	Dairy heifers (over 3 months)		127		
	Beef cows		128		
	Bulls from beef herd		129		
	Other beef cattle fat £ store £		130		
	Ewes and rams ewes £ rams £		131		
	Lambs: fat £ store £		133		
	Sows and boars: sows £ boars £		134		
	Other pigs (incl. in-pig gilts)		136		
	Poultry		137		Sub-Total
	Other livestock (specify)		138	139	
LIVESTOCK PRODUCE (include associated premiums and subsidies)	Milk		140		
	Cream, cheese, butter		141		
	Wool		142		
	Eggs		143		Sub-Total
	Other produce (specify)		144	145	
LIVESTOCK SUBSIDIES	Cattle e.g. hill cow		146		
	Sheep e.g. hill sheep		147		Sub-Total
	Other livestock grants e.g. brucellosis incentives		148	151	
CROP SALES AND SUBSIDIES	Wheat		152		
	Barley		153		
	Oats		154		
	Potatoes		155		
	Sugar beet		156		
	Other cash crops	Crop Code			
		157	(		
		158	(		
		159	(		
		160	(		
		161	(		
		409	(		
		410	(		
	Straw		162		
	Hay and silage		163		Sub-Total
	Feed roots		164	165	
SUNDRIES	Lime subsidy		166		
	Grant for keeping accounts		167		
	Other grants (exclude capital grants) specify:		168		
	e.g. FHDS guidance premium		169		
			170		
	Keep let		172		
	Contract work		173		
	Wayleaves, rents received etc.		174		Sub-Total
	Others (specify)		175	176	
NOTIONAL RECEIPTS	Rental value of house		177		
	Private use of car, electricity, etc.		178		
	Produce consumed: milk		179		
	poultry and eggs		180		Sub-Total
	other		181	182	
	Total Revenue (box 139+145+151+165+176+182)		183		
	Closing Valuation (box 52)		184		
	Total (box 183+184)		185		
	LOSS (box 119 less box 185)		186		

Valuations

LIVESTOCK (AT REALISTIC VALUES)	Item	OPENING VALUATION at (date)		CLOSING VALUATION at (date)	
		No.	£ Total	No.	£ Total
Dairy cows	1				
Bulls for dairy herd	2				
Dairy heifers	3				
Beef cows	6				
Bulls for beef herd	7				
Other beef cattle	8				
Ewes (incl. shearlings) and rams	11				
Other sheep	13				
Sows and boars	14				
Other pigs (incl. in-pig gilts)	16				
Laying hens (incl. breeders)	17				
Other poultry (specify)	18				
Other livestock (specify)	19				
Livestock products: wool	20				
Other livestock products (specify)	21				
(a) Total Livestock	22				

HARVESTED CROPS	Item	Tonne (if possible)	£	Tonne (if possible)	£
Wheat	23				
Barley	24				
Oats	25				
Potatoes	26				
Sugar beet	27				
Other cash crops	Crop Code 28				
	29				
	30				
	31				
	32				
	405				
	406				
Feed roots	33				
Other forage crops (specify)	34				
Hay (home grown)	35				
Silage (home grown)	36				
Straw (home grown)	37				
(b) Total Harvested Crops	39				

GROWING CROPS	Item				
Annual	407				
Perennial	408				
(c) Total Growing Crops	40				

GOODS IN STORE	Item				
Fertilisers	41				
Seeds and plants (purchased)	42				
Sprays	43				
Concentrates (purchased)	44				
Hay (purchased)	45				
Straw (purchased)	46				
Other purchased feed	47				
Fuel, oil, gas etc.	48				
Others (specify)	49				
(d) Total Goods in Store	51				
(e) Total Valuation (a) + (b) + (c) + (d)	52				

Appendix 2

TERMS AND DEFINITIONS USED IN FARM BUSINESS MANAGEMENT

Source: Ministry of Agriculture, Fisheries & Food.

1. The terms defined below relate in the main to financial transactions covering a period of twelve months; generally this is the accounting year, but when dealing with crops it is sometimes the harvest year, although the calculations can be made for any period if required. For cash flow purposes shorter periods of time, *e.g.*, one month or one quarter are often used.

VALUATIONS

2. The process of valuation is essentially one of estimation. The basis to be used may vary according to the purpose for which the valuation is made and alternative bases are given in some instances in the following section. The basis of valuation used in any data presented should be clearly stated and should be consistent throughout the period of any series of figures:

(a) *Saleable crops in store* are valued at estimated market value including deficiency payments less costs still to be incurred, *e.g.*, costs of marketing and storage. Market value and costs still to be incurred may be those either at the date of valuation or at the expected date of sale.

(b) *Growing crops* are valued at estimated cost up to the date of valuation. This may be either at variable costs or at estimated total cost. For most purposes, variable costs are preferable. Residual manurial values need to be taken into account only on change of occupancy.

(c) *Saleable crops ready for harvesting* but still in the ground should preferably be treated as 'Saleable crops in store' and should be valued as in 2(a) above less the estimated cost of harvesting. Alternatively, they may be treated as 'Growing crops'.

(d) *Fodder stocks (home-grown)* may be valued at estimated market value or variable costs. In calculating gross margins, valuation at variable costs is generally to be preferred. If market value is used, stocks of non-saleable crops, *e.g.*, silage, should be valued in relation to hay value adjusted according to quality. Fodder crops still in the ground, *e.g.*, kale, turnips, are treated as growing crops.

(e) *Stocks of purchased materials* (*including fodder*) are valued at cost net of discounts (where known) and subsidy (*e.g.*, fertiliser subsidy).

(f) *Machinery and equipment* are valued at original cost net of investment grants, less accumulated depreciation to date of valuation. Depreciation may be calculated by either the straight line or the reducing balance method.

(g) *Livestock*, whether for breeding, production or sale, are valued in their present condition at current market value, less cost of marketing. Fluctuations in market value which are expected to be temporary should be ignored.

OUTPUT TERMS

3. *Sales* are the value of goods sold for cash or on credit. In trading accounts sales exclude machinery and capital equipment. These items would, however, be included in capital accounts and cash flow calculations.

4. *Receipts* are monies received during the accounting period from the (see paragraph 3 above) sale of goods plus other remuneration, *e.g.*, subsidies, contracting, wayleaves. If practical, receipts should be recorded before deduction of off-farm marketing expenses such as commission and hire of containers.

5. *Revenue or [Income]* is Receipts adjusted for debtors, including outstanding Government subsidies such as deficiency payments, at the beginning and end of the accounting period.

6. *Returns* are Revenue adjusted for valuation changes.

Note: The value of goods and services produced on the farm for which no payment is made, *e.g.*, produce consumed in the farmhouse or supplied to workers, is not included in Returns but does form part of Gross Output.

7. *Gross Output* is total Returns plus the value of produce consumed in the farmhouse or supplied to workers for which no payment is made, less purchases of livestock, livestock products and other produce bought for resale. This can be calculated either for the farm as a whole or for separate sectors. Total Gross Output includes the Gross Output of livestock, crops and items of mis-

cellaneous output such as certain subsidies and grants, contracting, wayleaves.

8. *Gross Output of Livestock* (as a whole or for individual enterprises) is total Returns from livestock and livestock products less purchases of these items plus the value of any livestock produce consumed in the farmhouse, or supplied to workers for which no payment is made. Gross Output of Livestock includes production grants attributable to the enterprise, *e.g.*, hill sheep subsidy, calf subsidy.

9. *Gross Output of Crops* (as a whole or for individual enterprises) is total Returns from crops plus the value of any crop produce consumed in the farmhouse or supplied to workers for which no payment is made less the value of any produce bought for resale.

10. *Enterprise Output of a Livestock Enterprise** is its Gross Output plus the market value of livestock and livestock products transferred to another enterprise (transfers out) plus the market value of any production from the enterprise consumed on the farm less the market value of the livestock and livestock products transferred from another enterprise (transfers in).

11. *Enterprise Output of a Sale Crop Enterprise** is the total value of the crop produced, irrespective of its disposal; it equals Returns from the crop plus the market value of any part of the crop used on the farm. When this is calculated for the 'harvest year', as distinct from the accounting year, valuation changes are not relevant and the total yield of the crop is entered at market values plus deficiency payments.

12. *Enterprise Output from Forage* consists primarily of the sum of the Enterprise Outputs of grazing livestock. In addition it includes keep let and occasional sales *e.g.*, hay†, together with an adjustment for changes in the valuation of stocks of home grown fodder. *Note:* Changes in stocks caused by yield variations attributable to weather conditions, the severity or length of the winter or minor changes in livestock numbers or forage acres can be omitted from forage output and regarded as an item of miscellaenous output. Adjusted Enterprise Output from Forage is Enterprise Output from Forage less rented keep and purchase of bulk fodder.

---

* The Gross Output of an enterprise can be less than its Enterprise Output if any product of that enterprise is retained for use on the farm.

† Where sales of seed or fodder crops such as hay are a regular part of farm policy, they should be regarded as cash crops, not as Forage Crops.

13. *Net Output* is Gross Output less the cost of purchased feed, livestock keep, seed, bulbs, cuttings and plants for growing on.

14. *Standard Output* is the average Output (as defined in paragraphs 9–11) per acre of a crop or per head of livestock calculated as appropriate from national or local average price and yield data.

INPUT TERMS

15. *Purchases* are the value of materials and livestock acquired. In trading accounts purchases exclude machinery and capital equipment. These items would, however, be included in capital accounts and cash flow calculations.

16. *Payments* are monies paid during the accounting period for purchases (see paragraph 15 above) of material, livestock, and for services, off-farm marketing expenses, other owner-occupier expenses, interest and loan repayment. Non-cash items such as unpaid labour and estimated rental value should not be included.

17. *Expenditure* is Payments adjusted for creditors at the beginning and end of the year.

18. *Costs* are Expenditure with the following adjustments:
Add:
  (a) the opening valuation of the cost item;
  (b) the depreciation on items of capital expenditure including machinery;
  (c) to depreciation any loss made on the sale of machinery (*i.e.*, the difference between written down value and sale price);
  (d) to the cash wages of workers the value of payments-in-kind (if not already included in the earnings figure used);
Deduct:
  (a) the closing valuation of the cost item;
  (b) purchases of livestock, livestock products and any other produce bought for resale;
  (c) from depreciation any profit made on the sale of machinery (see c above);
  (d) from machinery costs any allowance for the private use of farm vehicles;
  (e) the value of purchased stores used in the farmhouse, (*e.g.*, coal, electricity), or sold off the farm, from the cost of that item;
  (f) the subsidies on fertiliser and lime (see paragraph 3(a)).

19. *Inputs* are Costs with the following adjustments made in order to put all farms on a similar basis for comparative purposes:

Add
  (a) the value of unpaid family labour, including the manual labour of the farmer and his wife;
  (b) for owner-occupiers and estimated rental value, less any cottage rents received and less an allowance for the rental value of the farmhouse;
Deduct:
  (a) any mortgage payments and other expenses of owner-occupation;
  (b) interest payments;
  (c) cost of paid management;
  (d) a proportion of the rental value of the farmhouse on tenanted farms.

20. *Fixed and Variable Costs for use in Gross Margin Calculations.* Variable Costs are defined as those costs which both can be readily allocated to a specific enterprise and will vary in approximately direct proportion to changes in the scale of that enterprise. The main Variable Costs are seed, fertilisers, sprays, concentrate feedingstuffs and much of the casual labour and contract machinery.

Fixed Costs are those costs which cannot readily be allocated to a specific enterprise and/or will not vary in direct proportion to small changes in the scale of the individual enterprises on the farm. Fixed Cost items include regular labour, machinery depreciation, rent and rates, and general overheads. Fuel and repairs are usually treated as Fixed Costs but glasshouse fuel is generally treated as a Variable Cost.

MARGIN TERMS

21. *Management and Investment Income* is the difference between Gross Output and Inputs. It represents the reward to management and the return on tenant's capital invested in the farm, whether borrowed or not.

22. *Net Farm Income* is Management Investment Income less paid management plus the value of the manual labour of the farmer and his wife. It represents the return to the farmer and his wife for their own manual labour and their management and interest on all farming capital, excluding land and buildings.

23. *Profit [or Loss]* is the difference between Gross Output and Costs. It represents the surplus or deficit before imputing any notional charges such as rental value or unpaid labour. In the accounts of owner-occupiers it includes any profit accruing from the ownership of the land.

24. *Gross Margin of a Crop Enterprise* is its Enterprise Output less its Variable Costs.

25. *Gross Margin of a Livestock Enterprise (non-land using)* is its Enterprise Output less its Variable Costs. (For barley beef, variable costs include those of any hay fed).

26. *Gross Margin from Forage* is Enterprise Output from Forage less (see paragraph 12):

(a) the Variable Costs directly attributable to individual types of grazing livestock, such as purchased concentrates, market value of home grown cereals fed, veterinary and medicine, AI fees;

(b) the Variable Cost of forage and catch crops, such as fertilizer, seed etc;

(c) the cost of purchased forage and rented keep.

27. *Gross Margin of a Grazing Livestock Enterprise* (*excluding Forage*) is its Enterprise Output less its Variable Costs.

28. *Gross Margin of a Grazing Livestock Enterprise* (*including Forage*) is the Gross Margin of a grazing livestock enterprise (excluding Forage) less the allocated Variable Costs of forage, the cost of purchased forage and rented keep where such an allocation is considered to be both possible and useful.

29. *Other Margin Terms.* When other uses of the term margin are made, they should be fully described on the basis of the definitions above.

OTHER TERMS

30. *Tenant's Capital* [*or Operating Capital*] is the estimated amount of capital on the farm, other than land and fixed equipment. There is no easy way of determining this sum precisely and estimates are made in several ways depending on the information available and the purpose for which the estimate is required. One method is to take the average of the opening and closing valuations of stores (feed, seed, fertilisers), machinery, crops and livestock; alternatively the closing valuation only may be taken. A third estimate is obtainable by calculating the annual average of several estimated valuations during the year. Whichever of these methods is used, the valuation may be at cost or at market value. Any estimate produced should be accompanied by a description of the method of calculation.

Appendix 3

CASE-STUDY FARM SCHEDULES

Schedule No.
1. Capital statement
2. Gross margin and fixed costs budget summary, year ending 31 March 1982
3. Dairy herd monitoring data—results, year ending 31 March 1981
4. Dairy herd monitoring data—target, year ending 31 March 1982
5. Grazing livestock gross margin budgets, year ending 31 March 1982
6. Grazing livestock gross margin targets, year ending 31 March 1982
7. Cash flow estimates, year ending 31 March 1982
8. Notes on cash flow estimates
9. Accumulative cash flow estimates/results, year ending 31 March 1982
10. Cash flow and revenue/expenditure results, year ending 31 March 1982
11. Trading account summary, year ending 31 March 1982
12. Gross margin accounts results summary compared to budget, year ending 31 March 1982
13. Fixed costs results, year ending 31 March 1982
14. Dairy herd gross margin, year ending 31 March 1982
15. Youngstock gross margin, year ending 31 March 1982

SCHEDULE 1. CAPITAL STATEMENT BUDGET

	Detail	£	Actual	£
1. Farming Assets				
(a) *Debts owed to a Business*				
Milk for previous month		14,500		
Other		Nil		
		14,500		
(b) *Saleable Crop Produce*				
Cereals				
Other crops				
		Nil		
(c) *Purchased Stores*				
Seeds		200		
Fertilisers		8,000		
Crop sprays & sundries		400		
Feedingstuffs	12 tonne	1,600		
		10,200		
(d) *Cultivations & Tenant Right*				
Cultivations		—		
Unused manurial values		5,000		
Tenant's fixtures		5,000		
		10,000		
(e) *Other Crop Produce*				
Hay	20 tonne	1,200		
Straw	10 tonne	200		
Silage	70 tonne	1,400		
		2,800		
(f) *Non-Breeding Livestock*				
In-calf heifers	42	14,700		
Yearling heifers	Nil	—		
Heifers over 3 months	32	4,800		
Calves	10	900		
		20,400		

Schedule 1—*continued*

	Detail	£	Actual	£
(g) *Breeding Livestock*				
Dairy cows	140	56,000		
Bulls	1 Hereford	600		
		56,600		
(h) *Machinery & Equipment*				
Tractors & vehicles		12,000		
Other field machinery		6,000		
Milking & fixed equipment		10,000		
		28,000		
TOTAL FARMING ASSETS*		142,000		
2. Farming Liabilities				
(a) Trade creditors		14,500		
(b) Bank overdraft		30,000		
(c) Other borrowed capital		Nil		
TOTAL LIABILITIES		44,500		
3. Net Farming Capital		98,000		

* Excluding land and buildings

SCHEDULE 2. GROSS MARGIN AND FIXED COSTS
BUDGET SUMMARY
(*Year Ending 31 March 1982*)

Gross Margins

	No.	*£* *Per* *head*	*Total* *£*
1. *Grazing Livestock*			
Dairy cows	140	470	65,800
Dairy replacements	85	115	9,800
TOTAL			75,600

	Hectares	*£* *Per* *hectare*	
2. *Cash Crops*			
All cash crops	Nil	—	—
All grazing livestock	80	945	75,600
TOTAL			

3. *Intensive L/Stk.*	*Per* *head*		

4. *Other Income*

5. *Notional Income*

	£ *Per* *hectare*	
TOTAL GROSS MARGIN	974	75,600
Less fixed costs	734	58,700
BUDGET PROFIT MARGIN	211	16,900

Add:	
Possible fixed cost savings	5,300
Target MoC less budget MoC	6,043
Target GM less budget GM	5,805

TARGET PROFIT MARGIN	34,048

Schedule 2—*continued*

Fixed Costs

	Total £	£ Per hectare
1. *Wages*		
Regular and N.I.		
Unallocated casual		
Wife	21,000	262·5
Paid management		
2. *Power & Machinery*		
Machine repairs	3,000	37·5
Contract/transport	2,400	30
Vehicle tax/insurance	400	5
Fuel and oil	2,800	35
Electricity & coal	2,400	30
Leasing charges	—	
Machinery depreciation	7,000	87·5
	18,000	225
3. *Property Charges*		
Rent	7,000	87·5
Rates	600	7·5
Depreciation:		
buildings & improvements	—	—
tenant's fixtures	600	7·5
Property repairs	2,800	35
	11,000	137·5
4. *Administration & Sundries*		
Water	600	7·5
General insurance	400	5
Office & telephone	600	7·5
Subscriptions & fees	400	5
Other	800	10
Lime	600	7·5
Overhead sprays	200	2·5
	3,600	45
5. *Finance Charges*		
Bank interest £30,000 @ 17 per cent	5,100	64
Mortgage interest	Nil	—
H.P. interest	Nil	—
TOTAL FIXED COSTS	58,700	734
Less		
Possible savings:		
Labour	2,000	25
Power & machinery	1,000	12·5
Property charges	400	5
Sundries	1,400	15
Finance	1,500	19
	5,300	66
TARGET FIXED COSTS	53,400	667

SCHEDULE 3. DAIRY HERD MONITORING DATA
(Results Year Ending 31 March 1981)

A. Monthly Data—Totals

Month	No. Cows* In milk	Dry	Total	No. calved† (H)	(C)	Milk sales Litre	£	Concs. tonne	Feed cost Other £	£	Total £	Margin over feed £
April	142	1	143	—		84,738	10,221	31		—	4,293	5,928
May	135	2	137	—		84,190	9,205	18		—	2,552	6,653
June	110	25	135	—		61,729	8,857	10		—	1,295	7,562
July	85	46	131	—		41,521	4,677	7		—	906	3,771
Aug	101	53	154	24	27	37,107	4,331	5		—	645	3,686
Sept	107	51	158	6	19	57,461	7,651	21		—	2,908	4,743
Oct	107	41	148		28	76,015	9,840	35		—	4,848	4,992
Nov	119	29	148		12	85,282	10,990	40		—	5,560	5,430
Dec	133	10	143		19	105,093	13,565	50		—	7,000	6,565
Jan	142	0	142		10	118,055	15,406	55		—	7,700	7,706
Feb	142	0	142			107,337	13,954	50		—	7,000	6,954
March	140	0	140			116,340	15,124	54		—	7,560	7,564
YEAR	‡	‡	143.4‡	30	115	974,868	123,821	376		Nil	52,267	71,554

* At start of month † Include purchases ‡ Average no. per year
H = heifers C = cows

Schedule 3—*continued*

| | B. Monthly per litre | | | C. Monthly data per Cow | | | D. Accumulative Data | |
| | | | | | | | Per cow | |
Month	Sale price p	Feed cost p	Margin p	Margin over feed £	Milk sales litre	Total margin £	Margin £	Milk sales litre
April	11.79	5.07	6.72	41	593	5,928	41	593
May	10.96	3.15	7.81	48	614	12,581	89	1,207
June	11.175 *14.35	2.10	12.25	56	457	20,143	145	1,664
July	11.560	2.18	9.38	29	317	23,914	174	1,981
Aug	11.996	1.77	10.22	24	243	27,600	198	2,224
Sept	12.570 *13.316	5.06	8.26	30	364	32,343	228	2,588
Oct	12.94	6.38	6.56	33	514	37,335	261	3,102
Nov	12.89	6.52	6.37	37	576	42,765	298	3,678
Dec	12.91	6.66	6.25	46	735	49,330	342	4,413
Jan	*13.05	6.52	6.48	54	831	57,036	396	5,244
Feb	13.00	6.48	6.52	49	755	63,990	445	5,999
March	13.00	6.50	6.50	54	831	71,554	500	6,800
YEAR	12.70	5.36	7.34					

* Includes additional payment for previous months.

SCHEDULE 4. DAIRY HERD MONITORING DATA
(Target Budget Year Ending 31 March 1982)

						A. Monthly Data—Totals					
	No. cows*			No. calved†		Milk sales			Feed cost			Margin over feed
Month	In milk	Dry	Total	(H)	(C)	Litre	£	Concs. tonne	Other £	£	Total £	£
April			140									9,100
May			140									9,100
June			135									7,425
July			130									5,720
Aug			130									4,550
Sept			140	20	40							5,040
Oct			152	12	25							5,472
Nov			150		20							6,150
Dec			145		15							7,540
Jan			142		10							8,946
Feb			140									7,980
March			140									8,820
YEAR	‡	‡	‡	32	110							85,843

* A start of month   † Include purchases   ‡ Average no. per year

Schedule 4—*continued*

| | B. Monthly per litre | | | C. Monthly Data per Cow | | D. Accumulative Data | | |
| | | | | | | | Per cow | |
Month	Sale price p	Feed cost p	Margin p	Margin over feed £	Milk sales litre	Total margin £	Margin £	Milk sales litre
April	12.7	4.0	8.7	65	750	9,100	65	750
May	11.8	2.0	9.8	65	670	18,200	130	1,420
June	11.8	1.6	10.2	55	540	25,625	185	1,960
July	12.6	1.6	11.0	44	400	31,345	229	2,360
Aug	13.0	1.4	11.6	35	300	35,895	264	2,660
Sept	13.5	5.0	8.5	30	360	40,935	294	2,020
Oct	14.0	7.0	7.0	36	520	46,407	330	3,540
Nov	14.0	7.0	7.0	41	580	52,557	371	4,120
Dec	14.0	7.0	7.0	52	740	60,097	423	4,860
Jan	14.0	6.5	7.5	63	840	69,043	486	5,700
Feb	14.0	6.5	7.5	57	760	77,023	543	6,460
Mar	14.0	6.5	7.5	63	840	85,843	606	7,300
YEAR			8.3		7,300			

SCHEDULE 5. GRAZING LIVESTOCK GROSS MARGIN BUDGETS
(Year Ending 31 March 1982)

Average no. cows 140 replacements 85	Dairy herd		Per cow	Dairy replacements		TOTAL	Cash flow
Closing Valuation	(No)	£	£	(No) ditto	£ Opening valuation	£	
Total valuation	140	56,000	400	85	21,000	77,000	
Stock Sales							
Cull cows/heifers	30	10,500		2	800	11,300	11,300
Surplus heifers	– –	– – – –		8	4,200	4,200	4,200
Calves	100	4,700		—	– – – – –	4,700	4,700
Casualties	2	– – – –	——	2	– – – – –	– – – –	– – – –
Produce sales		132,300	945			132,300	132,300
Transfers out	44	2,200	——	32	16,000	18,200	
Sub-total (A)		205,700		129	42,000	247,700	152,500
Opening valuation							
				1 Bull	500		
				42 Ic	14,700		
				32<3m	4,800		
				10<3m	900		
Total valuation	140	56,000	400	85	21,000	77,000	
Stock Purchases							
Transfers in	32	16,000		44	2,200	18,200	——
Sub-total (B)		72,000		129	23,200	95,200	
Gross output (A−B)		133,700	955	221*	18,800	152,500	
Variable Costs			Per head				
Bought concentrates		52,500	375	60	5,100	57,600	57,600
Other bought feed		– – – –	– –				
Homegrown grain		– – – –	– –				
Vet & medicines		3,500	25	6	500	4,000	4,000
Other items		2,100	15	5	400	2,500	2,500
Total excluding forage		58,100	415	71	6,000	64,100	64,100
Gross margin excluding forage		75,600	540	150	12,800	88,400	
Forage costs		9,800	70	35	3,000	12,800	12,800
GROSS MARGIN		65,800	470	115	9,800	75,600	
Milk yield per cow (litres)			7,000				
Milk price per litre (*p*)			13.50				
Margin over concs per cow (£)			570				

* Enter gross output per head (total ÷ ave no.)

SCHEDULE 6. GRAZING LIVESTOCK GROSS MARGIN TARGETS
(Year Ending 31 March 1982)

Average no. cows 140 replacements 85	Dairy herd		Per cow	Dairy replacements		TOTAL	Cash flow
Closing valuation	(No)	£	£	(No)	£	£	
				ditto	Opening valuation		
Total valuation	140	56,000	400	85	21,000	77,000	
Stock Sales							
Culls cows/heifers	28	9,800		1	400	10,200	
Surplus heifers	– –	– – – –		13	7,150	7,150	
Calves	104	5,400		2	– – – – –	5,400	
Casualties	Nil	– – – –		– – –	– – – – –	132,300	
Produce sales		132,300	945				
Transfers out	44	2,200		28	14,000	16,200	
Sub-total (A)		205,700		129	42,500	248,250	
Opening valuation							
				1 Bull	500		
				42 Ic	14,700		
				32>3m	4,800		
				10<3m	900		
Total valuation	140	56,000	400	85	21,000	77,000	
Stock Purchases							
Transfers in	28	14,000		44	2,200	16,200	———
Sut-total (B)		70,000		129	23,200	93,200	
Gross output (A−B)		135,700	969	228*	19,350	155,050	
Variable Costs				Per head			
Bought concentrates		52,500	375	55	4,675	57,175	
Other bought feed				– – –			
Homegrown grain				– – –			
Vet & medicines		3,124	22	6	500	3,624	
Other items		1,846	13	5	400	2,246	
Total excluding forage		57,470	410	66	5,575	63,045	
Gross margin excluding forage		78,230	559	162	13,775	92,005	
Forage costs		9,080	64	32	2,720	11,800	
GROSS MARGIN		69,150	495	130	11,055	80,205	
Milk yield per cow (litres)			7,000				
Milk price per litre (p)			13.5				
Margin over concs per cow (£)			570				

* Enter gross output per head (total ÷ ave no.)

SCHEDULE 7. CASH FLOW ESTIMATES
(Year Ending 31 March 1982)

Details		Total	April to June 1981	July to Sept 1981	Oct to Dec 1981	Jan to March 1982
Trading Items						
Cull cows/heifers	1	11,300	1,800	6,400	2,400	700
Surplus heifers	2	4,200	—	—	4,200	—
Calves	3	4,700	—	—	3,300	1,400
Milk	4	132,300	37,000	19,800	30,500	45,000
	5					
	6					
	7					
	8					
	9					
	10					
	11					
	12					
	13					
Total Trading	14	152,500	38,800	26,200	40,400	47,100
Capital Items						
M/c & Equipment	15					
Grants	16					
Loans	17					
	18					
	19					
VAT	20					
(A) TOTAL RECEIPTS	21	152,500	38,800	26,200	40,400	47,100
Trading Items						
Livestock Purchases	22	Nil	—	—	—	—
Concentrates	23	57,600	13,600	6,000	15,500	22,500
Other feed	24	Nil	—	—	—	—
Vet & medicines	25	4,000	1,000	1,000	1,000	1,000
Livestock sundries	26	2,500	600	600	700	600
Fertilisers & lime	27	10,600	—	—	—	10,600
Seeds & crop sundries	28	3,000	1,500	1,500	—	—
Regular labour	29	21,000	5,000	5,000	5,000	6,000
Contract	30	2,400	—	2,400	—	—
M/c repairs & tax/ins.	31	3,400	800	900	800	900
Fuel & power	32	5,200	1,300	1,300	1,300	1,300
Rent & rates	33	7,600	—	3,800	—	3,800
Property repairs	34	2,800	700	700	700	700
Misc. inc. Insurance	35	2,800	700	700	700	700
Interest charges	36	5,100	1,400	1,400	1,200	1,100
Total trading	37	128,000	26,600	25,300	26,900	49,200
Capital and Private Items						
Private drawings & tax	38	6,800	2,200	1,200	2,200	1,200
M/c & equipment	39	15,500	5,500	—	—	10,000
Land & improvements	40	—	—	—	—	—
Loan repayments	41	—	—	—	—	—
	42					
	43					
VAT	44					
(B) TOTAL PAYMENTS	45	150,300	34,300	26,500	29,100	60,400
(C) SURPLUS/DEFICIT	46	+2,200	+4,500	(300)	+11,300	(13,300)
Opening bank balance (overdraft)	47		(30,000)	(25,500)	(25,800)	(14,500)
Closing bank balance (overdraft)	48		(25,500)	(25,800)	(14,500)	(27,800)

SCHEDULE 8. NOTES ON CASH FLOW ESTIMATES
(Year Ending 31 March 1982)

1. Based on budget data [not on target data].
2. *Cull cows and heifers.* 5 first quarter, 16 second + 2 heifers, 7 third, 2 fourth.
3. *Surplus heifers.* All in third quarter.
4. *Calves.* Some may be sold in September *but* all shown in third quarter, except 20 sold in fourth quarter.
5. *Milk.* Distribution is assumed to be the same as in 1980–1:

Quarter	Sales	Per cent		£
1	(March, April & May)	28	£132,300 =	37,000
2	(June, July & Aug)	15	£132,300 =	19,800
3	(Sept, Oct & Nov)	23	£132,300 =	30,500
4	(Dec, Jan & Feb)	34	£132,300 =	45,000
		100		132,300

6. *Concentrates.* Payments are assumed to be 1 month in arrears of deliveries, *i.e.*, payment in April–June quarter is for feed delivered in March–May.
7. *Veterinary medicines & livestock sundries.* Evenly distributed throughout the year.
8. *Fertilisers.* Current year's supply has been paid for prior to start of year—payments in this year are for the next year's supply and this is budgeted in the fourth quarter.
9. *Seeds and crop sundries.* First two quarters.
10. *Contract.* Paid in second quarter.
11. *Rent.* Half year's payment due in arrear in September and March.
12. *Remaining fixed cost items.* All evenly distributed throughout the year.
13. *Private drawings and tax.* Drawings £1,200 per quarter + £1,000 tax in first and third quarters.
14. *Machinery and equipment.* The outlay on a tractor (£3,500) and farm vehicle (£2,000) are definite commitments. It is also hoped to pay £10,000 in the 4th quarter representing half the cost of completely updating the parlour, the remaining £10,000 for building works being funded in the next financial year. But this work has not yet been given the go ahead.

SCHEDULE 9. ACCUMULATIVE CASH FLOW
ESTIMATES/RESULTS
(Year Ending 31 March 1982)

Details		A	B	C	D	E	F	G	H
		Three Months		*Six Months*		*Nine Months*		*Twelve Months*	*Revised*
		Budget	*Actual*	*Budget*	*Actual*	*Budget*	*Actual*	*Budget*	*estimate*
Trading Items									
Cull cows/heifers	1	1,800	2,200	8,200	8,000	10,000	9,600	11,300	(1,000)
Surplus heifers	2	—	—	—		4,200	6,500	4,200	+2,300
Calves	3	—	—	—		3,300	4,000	4,700	+1,000
Milk	4	37,000	36,500	56,800	60,000	87,300	93,000	132,300	+7,700
	5								
	6								
	7								
	8								
	9								
	10								
	11								
	12								
	13								
Total Trading	14	38,800	38,700	65,000	68,000	105,400	113,100	152,500	+10,000
Capital Items									
M/c & equipment	15								
Grants	16								
New loans	17								
	18								
	19								
VAT	20								
(A) TOTAL RECEIPTS	21	38,800	38,700	65,000	68,000	105,400	113,100	152,500	+10,000 =162,500

Schedule 9—continued

Details		A	B	C	D	E	F	G	H
		Three Months		Six Months		Nine Months		Twelve Months	Revised
		Budget	Actual	Budget	Actual	Budget	Actual	Budget	estimate
Trading Items									
Livestock purchases	22	Nil	Nil	Nil	Nil	Nil	600	Nil	+600
Concentrates	23	13,600	19,600	25,600	26,000	41,100	41,500	63,600	+400
Other feed	24	—	—	—	—	—	1,000	—	+1,000
Vet, medicines	25	1,000	1,000	2,000	2,100	3,000	3,200	4,000	+400
Livestock sundries	26	600	600	1,200	1,200	1,900	2,100	2,500	+300
Fertilisers & lime	27	—	—	—	—	—	1,400	10,600	+2,400
Seeds & crop sundries	28	1,500	1,500	3,000	2,500	3,000	3,400	3,000	+600
Regular labour	29	5,000	5,000	10,000	10,200	15,000	15,500	21,000	+1,000
Contract	30	—	—	2,400	2,400	2,400	2,400	2,400	Ditto
M/c repairs/contract	31	800	800	1,700	1,800	2,500	2,700	3,400	+400
Fuel & power	32	1,300	1,100	2,600	2,800	3,900	4,200	5,200	+400
Rent & rates	33	—	3,500	7,300	7,300	7,300	7,300	11,100	Ditto
Propery repairs	34	700	500	1,400	1,500	2,100	1,600	2,800	(600)
Misc. inc. insurance	35	700	700	1,400	1,600	2,100	2,200	2,800	+200
Interest charges	36	1,400	1,300	2,800	2,600	4,000	3,500	5,100	(600)
Total Trading	37	26,600	35,600	61,400	62,000	88,300	92,600	137,500	+6,500
Capital and Private Items									
Private drawing & tax	38	2,200	2,500	3,400	3,800	5,600	6,800	6,800	+1,200
M/c & equipment	39	5,500	6,000	5,500	6,000	5,500	6,500	15,500	+1,000
Land & improvements	40								
Loan repayments	41								
	42								
	43								
VAT	44								
(B) TOTAL PAYMENTS	45	34,300	44,100	70,300	71,800	99,400	105,900	159,800	+8,700 = 168,500
(C) SURPLUS/ DEFICIT	46	+4,500	(5,400)	(5,300)	(3,800)	+6,000	+7,200	(7,300)	(6,000)
Opening bank bal. (o/draft)	47	(30,000)	(24,000)	(24,000)	(24,000)	(24,000)	(24,000)	(24,000)	(24,000)
Closing bank bal. (o/draft)	48	(25,500)	(29,400)	(29,300)	(27,800)	(18,000)	(16,800)	(31,300)	(30,000)

SCHEDULE 10. CASH FLOW AND REVENUE/ EXPENDITURE RESULTS
(Year Ended 31 March 1982)

Details		Actual cash flow	Add closing debtors & creditors	Subtract opening debtors & creditors	Actual revenue & expenditure	Budget revenue & expenditure	Variance
Trading Items							
Cull cows/heifers	1	10,300	Nil	Nil	10,300	11,300	(1,000)
Surplus heifers	2	6,500	Nil	Nil	6,500	4,200	+2,300
Calves	3	5,700	Nil	Nil	5,700	4,700	+1,000
Milk	4	140,000	14,000	13,000	141,000	132,300	+8,700
	5						
	6						
	7						
	8						
	9						
	10						
	11						
	12						
	13						
Total Trading	14	162,500	14,000	13,000	163,500	152,500	+11,000
Capital Items							
M/c & equipment	15						
Grants	16						
New loans	17						
	18						
	19						
VAT	20						
(A) TOTAL RECEIPTS	21	162,500	14,000	13,000	163,500	152,500	+11,000
Trading Items							
Livestock purchases (bull)	22	600	—	—	600	—	+600
Concentrate feed	23	64,000	5,000	10,000	59,000	57,600	+1,400
Other feed	24	1,000	—	—	1,000	Nil	+1,000
Vet & medicines	25	4,400	400	600	4,200	4,000	+200
Livestock sundries	26	2,800	400	500	2,700	2,500	+200
Fertilisers & lime	27	13,000	—	—	13,000	10,600	+2,400
Seeds & crop sundries	28	3,600	—	600	3,000	3,000	Ditto
Regular labour 1	29	22,000	800	1,000	21,800	21,000	+800
Contract	30	2,400	—	—	2,400	2,400	Ditto
M/c repairs & tax/ins.	31	3,800	800	600	4,000	3,400	+600
Fuel & power	32	5,600	600	600	5,600	5,200	+400
Rent & rates	33	11,100	—	3,500	7,600	7,600	Ditto
Property repairs	34	2,200	—	200	2,000	2,800	(800)
Misc. inc. insurance	35	3,000	600	400	3,200	2,800	+400
Interest charges	36	4,500	—	—	4,500	5,100	(600)
Total trading	37	144,000	8,600	18,000	134,600	128,000	+6,600
Capital and Private Items							
Private drawings & tax	38	8,000	—	—	8,000	6,800	+1,200
M/c & equipment	39	16,500	—	—	16,500	15,500	+1,000
Land & improvements	40						
Loan repayments	41						
	42						
	43						
VAT	44						
(B) TOTAL PAYMENTS	45	168,500	8,600	18,000	159,100	150,300	+8,800
(C) SURPLUS/DEFICIT	46	(6,000)			+4,400	+2,200	+2,200
Opening bank balance (overdraft)	47	(24,000)				(30,000)	
Closing bank balance (overdraft)	48	(30,000)				(27,800)	

SCHEDULE 11. TRADING ACCOUNT SUMMARY
(Year Ended 31 March 1982)

Opening valuation	£	£	Closing valuation	£	£
Livestock			*Livestock*		
Dairy cows	56,000		Dairy cows	56,400	
Bull	600		Bulls	600	
Dairy replacements	20,400	77,000	Dairy replacements	21,100	78,100
Crop Produce			*Crop Produce*		
Hay	1,200		Hay	1,200	
Straw	200		Straw	200	
Silage	1,400	2,800	Silage	2,400	3,800
Seeds	200		Seeds	200	
Fertilisers	8,000		Fertilisers	9,200	
Crop sprays & sundries	400		Crop sprays & sundries	400	
Feeding stuffs	1,600	10,200	Feeding stuffs	1,600	11,400
TOTAL		90,000	TOTAL		93,300
Trading Expenditure			*Trading Revenue*		
Livestock purchases		600	Cull cows/heifers		10,300
Feeding stuffs		60,000	Surplus heifers		6,500
Livestock sundries		2,700	Calves		5,700
Vet & medicines		4,200	Milk		141,000
Fertilisers and lime		13,000			
Seeds and crop sundries		3,000			
Wages & national insurance		21,800			
Power and machinery inc. contract		12,000			
Rent and rates		7,600			
Property repairs		2,000			
Sundries		3,200			
Interest charges		4,500			
TOTAL		134,600	TOTAL		163,500
Depreciation			*Notional Income*		
Machinery and equipment		7,000	Produce consumed		400
Tentant's fixtures and improvements		600	R.V. house. Use of car etc.,		1,200
TOTAL		7,600	TOTAL		1,600
PROFIT		26,200	LOSS		—
		258,400			258,400

SCHEDULE 12. GROSS MARGIN ACCOUNT RESULTS SUMMARY COMPARED TO BUDGETS
(Year Ending 31 March 1982)

Gross Margins	Results			Budget		
	Total £	*Stock no.*	*£ per head*	*Total* £	*Stock no.*	*£ per head*
1. Grazing livestock						
		Hectares	Per hectare		Hectares	Per hectare
Total (A)						
2. Cash crops						
Total (B)						
3. All land-using enterprises (A+B)						
4. Intensive livestock						
5. Other income						
6. Notional income			Per hectare			Per hectare
TOTAL GROSS MARGIN Less fixed costs (−)						
PROFIT MARGIN						

SCHEDULE 13. FIXED COSTS
(Year Ending 31 March 1982)

	Total £	Result Per hectare £	Budget £	Result previous year £
Wages				
Regulars & national insurance				
Unallocated casual labour				
Wife's wages				
Paid management				
Secretarial				
Sub total				
Power and Machinery				
Machinery & vehicle repairs				
Fuel & oil				
Contract & transport				
Leasing				
Electricity & coal				
Machinery depreciation				
Vehicle tax & insurance				
Sub total				
Property Charges				
Rent				
Rates				
Depreciation: Bldgs & improvements				
Tenant's fixtures				
Property repairs				
Sub total				
Sundries				
Water charges				
General insurance				
Office, telephone, professional fees				
Others				
Sub total				
Financial Charges				
Bank charges & interest				
Other interest charges				
Sub total				
TOTAL				

SCHEDULE 14. DAIRY HERD GROSS MARGIN
(Year Ending 31 March 1982)

Average number cows			Results		Budget	
	.........Actual					
	.........budget		Total	Per cow	Total	Per cow
Output	No.	Per head	£	£	£	£
Closing valuation	.........cows					
Stock sales	.........cows					
	.........calves					
Milk sales	.........litres					
Transfers out	.........calves					
Sub-total (A)						
Opening valuation	.........cows					
Purchases	....					
Transfers in	.........heifers					
Sub-total (B)						
GROSS OUTPUT (A−B)						
Variable Costs		Per tonne				
Purchased concs.	.........tonne					
Other purch. feed	.........tonne					
	.........tonne					
Home-grown grain	.........tonne					
Vet & medicines						
Others						
Total (excl. forage)						
Gross Margin (excl. forage)						
Forage costs.........hectares						
Gross Margin (excl. unit value appn.)						
Appreciation in cow unit values						
Gross margin (incl. unit value appn.)						
GROSS MARGIN PER HECTARE						
Margin over feed cost per cow						
Milk price per litre						
Feed cost per litre*						
Margin over feed cost per litre*						
Milk yield per cow (litres)						
Calf output per cow						
Herd maintenance per cow						

* Including homegrown grain & other purchased feed

SCHEDULE 15. YOUNGSTOCK GROSS MARGIN
(Year Ending 31 March 1982)

		Result		Budget	
Average number.........actual			*Per*		*Per*
.........*budget*		*Total*	*head*	*Total*	*head*
Output	*Detail*	£	£	£	£
Closing valuation					
Stock sales					
Produce sales					
Transfers out					
Subsidies					
Sub-total (A)					
Opening valuation					
Stock purchases					
Transfers in					
Sub-total (B)					
Gross output (A−B)					
Variable Costs					
Bought concentrates					
Other bought feed					
Home-grown grain					
Vet. & medicines					
Other items					
Total excluding forage					
Gross margin excluding forage					
Forage costs.........					
GROSS MARGIN					

INDEX

Accounts—example, 39
—, standardisation and definition of terms, 32–3, 36
—, use for business analysis, 37, 129
Agricultural Mortgage Corporation, 172–4
Appreciation in livestock values, 57, 82–3
Assets and liabilities, statement, 49

Balance sheet, 47–9
—, variations between farms, 60–1
Beef cattle, margins from, 106–9
Beef prices—and calf prices, 79–80
—, and profitability, 63
Borrowed capital—ability to repay, 168–9, 174–5
—, sources, 171–2
Breakeven budgets, 168–9
Breeding records, 164–5
Brinkmanship recording system, 152–61
Budgeting—case study example, 131, 141–9
—, principles, 116–18, 121–2
Buildings, assessment, 126, 131–2
Business approach, need for, 15
Business strategy, 17

Calf prices, 79–80
Calf rearing, costs of, 105–6
Capital—assessment, 129–30
—, classification of, 166–8
—, requirements, 60–1, 132–3, 168–78

—, return on, 138–9, 169–70
—, sources of, 171–2
Cash flow—budgeting and control, 146–9
—, dairy replacements, 94
Cereals on the dairy farm, 101–3
Channel Island producers, 75–6
Communications, 19
Comparative analysis, 44, 112–14, 142
Competitive enterprises, 26
Complementary enterprises, 26
Complete enterprise costings, 118
Compositional quality, and milk prices, 73–5
Concentrates—costs, 68–9
—, use per litre milk, 162
Costs, definitions, 190
Credit sources, 171–2
Creditors, 33
Cull cow prices, 81

Dairy heifers (replacements)—
—, age at calving, 91–4, 133
—, place on the farm, 88–9
—, profitable systems, 94–6
Debtors, 33
Decision-making, 16–17, 23, 30
Delegation, 18
Depreciation, machinery, 35
Diminishing returns, 27
Diversification—reasons for, 25
—, and role of dairy replacements, 88–9

Economic objectives, 23
Enterprise management efficiency, 50–1, 59
Equi-marginal returns, 29–30

215

Expenditure, definitions, 33, 190

Feed control, 143–4, 161–2
—, records, 44, 151
Fertiliser costs, 132, 138–9
Finance (interest) charges, 46, 52–3
—, as available costs, 144–5
Fixed costs, 37, 41, 45–7, 50–1, 191
—, budgeting, 135–6
—, limitations in use for comparative analysis, 114
Forage costs—allocation, 43, 83–4
—, dairy replacements, 89–90
Forage strategy, 133–4, 143–5
Functions of management, 16, 20

Grassland—efficiency, 118–20
—, milk from, 160–1
Gross margins, 37, 50–1, 114–16, 192
—, budgets, 135–40
—, calculations, 40, 42–3
—, factors affecting, 65–72, 138
—, grassland efficiency, 118–21
—, standards, 55–8
—, variations in 59–60, 61–2, 138–40
Gross output, 37, 40, 188–9

Hedge against inflation, 34, 58, 82–3
Herd depreciation, 80–1
Husbandry, efficiency measures, 162–4

Income definitions, 188
Inflation, 16, 34
Interest (finance) charges, 46, 52–3
—, budget estimates, 136, 168

Job descriptions, 21
Job opportunities, 15, 180–1

Labour costs, 53–4, 135–6
—, and farm policy, 127, 131, 133–4
—, productivity, 56–7
Lactation curves, 152–3
—, monitoring, 154–61
Land, assessment, 125
—, utilisation, 133–4
—, values, 170–1
Liquidity, 48
Livestock units and allocation of forage costs, 43, 89
Loan repayment methods, 172–3

Machinery and equipment inventory, 127–8
Machinery and power costs, 55–6, 135
Machinery and productivity, 56–8
Management and Investment Income, 37, 191
Marketing, 19, 86–7
Margin over Concentrates, 65–6, 71–3, 132–3, 137–8, 141
Milk prices—and compositional quality, 19, 73–5
—, and EEC policy, 86
—, liquid and manufactured, 84–5
—, relative to feed costs, 64
—, relative to wages, 24
—, seasonality of production, 77
Milk yield—and MoC data, 67–8
—, monitoring, 137, 151

Net capital (worth), 47–9
Net Farm Income, 36, 191
Non-land using enterprises, 110
Number cows, 25
—, milk producers, 16, 27

Objectives, managed by, 122
Opportunity costs, 28–9, 102, 118–21

Partial budgeting, 116–18, 144–5
Payments, definitions, 190
Pedigree breeding, 97–9

Potatoes, 104
Power and machinery costs, 55, 135
Productivity, 56
Profit definitions, 36–7, 41, 191
Profitability—dairy replacements, 89–91
—, relative to beef prices, 63
—, size of farm, 176–7
—, variation according to efficiency, 138–40
—, variation between farms, 57, 59, 61–2
—, variation from year to year, 63–4
Purchases, definitions, 190

Receipts, definitions, 188
Records—need for, 44, 151
—, feed, 161–2
—, breeding and veterinary, 164
Rent equivalent, 51, 173
Resource assessment, 125–30, 131–3
Revenue, definition, 31, 188

Sales, definition, 188
Seasonality of milk prices, 77
Share farming, 179

Sheep, 110
Silage—milk from, 158–60, 163
—, strategy, 133–4
Specialisation, 15, 19, 24
Substitution—principle of, 29
—, feeding dairy cows, 69–70
—, use of homegrown cereals, 102–3
Supplementary enterprises, 26
Sugar beet, 104
System efficiency, 50–1, 59–60

Tenant right, 127–8
Trading account, 39, 184–5

Valuations, 33–4, 187–8
—, buildings and equipment, 53
—, standards per hectare, 58
Variable costs, 37, 40, 45–7, 191
—, allocation, 43–4
Veterinary records, 164–5

Wages—relative to milk prices, 24, 53
—, 1975–1981, 53

Yield, see milk yield